CASTING DEFECTS HANDBOOK: IRON & STEEL

George Goodrich
Technical Editor

AFS

Published and distributed
by the

American Foundry Society
Schaumburg, Illinois 60173
www.afsinc.org

Fifth Printing 2019

ISBN 978-0-87433-314-5

Printed in the United States of America

Edited by Susan P. Thomas/AFS

Foreword

Since publication of the first edition of the AFS *Casting Defects Handbook*, it has been considered the go-to pocket reference for metalcasting personnel worldwide for more than 35 years. This edition has been rewritten under the direction of the American Foundry Society (AFS) Cast Iron Division (Division 5) and the Steel Division (Division 9) to provide the metalcasting industry with updates to reflect the changes in foundry technology.

The original handbook was written by the Molding Methods and Materials Group Special Publications Committee 80-G, Section 2, Casting Defects. While AFS no longer has a committee dedicated exclusively to casting defects, this matter is addressed by the appropriate technical committees.

In 2005, it was decided to review and update the book to include the many technological changes that have occurred since the book was first introduced. The purpose of the book remains the same, to provide foundries with the updated information necessary to reduce casting defects and increase productivity to stay competitive in today's market.

During the interval when this book was first published and today, many changes have occurred that were not reflected in the original version. Flaskless molding, vertical molding, high density molding with automatic molding machines, no bake molding, and the preponderance of electric melting vs. cupola melting are just a few examples. In addition, the mechanism of identifying the defect cause has an entirely new discipline, lean six sigma. Even the mix of metal importance has changed. For example, malleable iron usage is declining contrasting with the increased popularity of ductile iron.

Much of the original content has been retained because it continues to be of value today. This edition combines all of the strengths of the original handbook as well as incorporating the latest industry updates with a focus on ferrous alloys. The text is formatted in an easy-to-use outline style supplemented with figures illustrating actual casting defects.

George M. Goodrich
Technical Editor

Acknowledgements

The American Foundry Society gratefully acknowledges the following individuals and their companies for their time and efforts used to update this handbook:

Phil Bruno, *Fiat Group CNH*

Bryan Burton, *ThyssenKrupp Waupaca*

Prabhaker Gangasani, *Wells Manufacturing Co., Dura-Bar Division*

George M. Goodrich, *Stork Climax Research Services*

Kathy Hayrynen, *Applied Process*

Chantal Labrecque, *Rio Tinto Iron & Titanium Co.*

Ryan Leitermann, *Metal Technologies West Allis Ductile Iron Plant*

Fred Linebarger, *Miller & Co.*

Greg Miskinis, *ThyssenKrupp Waupaca*

Eric Nelson, *The Dotson Company Inc.*

Richard Obara, *Emerson Climate Technologies*

Von L. Richards, *Missouri University of Science and Technology*

Hathibelagal Roshan, *Maynard Steel*

Steve Sauer, *Carpenter Brothers, Inc.*

Phil Seaton, *Chrysler*

Brad Steinkamp, *Wells Manufacturing Co., Dura-Bar Division*

David Tucker, *Caterpillar*

Gerald A. Wilkinson, *Kohler Co.*

Leo Baran, *AFS*

Scott Lammers, *AFS*

Mark Nagel, *AFS*

Henry Parrilli, *AFS*

Susan P. Thomas, *AFS*

Contents

CHAPTER 1

Broken or Cracked Castings

Description

This type of defect refers to castings which have been broken or cracked by mechanical action, rough handling or thermal shock.

Causes

1. Casting and Pattern Design

- Irregular sections, isolated heavy sections, light sections or projections that require careful handling during shakeout, cleaning, finishing and heat treating
- Lack of fillets
- Lack of proper reinforcing ribs, tie bars or stress relieving

2. Pattern Equipment

- Failure to provide proper break-off notches on the gates and risers

3. Flask Equipment and Handling Devices

- Flask bars located too close to the riser or sprue
- Flask bars extend into deep pockets, prohibiting normal collapse of the sand

4. Gating and Risering

- Lack of adequate fillets
- Incorrectly placed or dimensioned cracking strips and tie bars
- Gating system design promotes stress or segregation due to turbulence

Fig. 1-1. An early rough shakeout of this gray iron casting, while hot and weak, resulted in broken gates.

5. Molding Sand

- Poor sand collapsibility due to excessive hot or dry compressive strength
- Low sintering point materials in the sand
- High hot compressive strength coupled with hot sand deformation

6. Core Practice

- Poor collapsibility due to excessive hot compression strength
- Low hot deformation when coupled with poor collapsibility
- Rods or arbors located too close to the surface of cores
- Over-reinforced cores

7. Molding Practice

- Mold rammed too hard
- Rods, arbors and gaggers located too close to the mold surface
- Risers or sprues located too close to a flask bar
- Excessive swabbing or sponging
- Improper use of chills or excessive chilling

Fig. 1-2. Poor design of the cracking strip increased the breakage of this malleable casting in the hard iron condition.

8. Metal Composition

- Improper metal composition for the application
- Metal composition with too high shrinkage characteristic
- White Iron composition requires special handling during shakeout

9. Melting Practice

- Excessive carbide stabilizers promote high shrinkage
- Composition possessing a very high shrinkage characteristic
- Gaseous or over oxidized metal
- Contamination

Fig. 1-3. Careless finning caused this casting to break in the gate area.

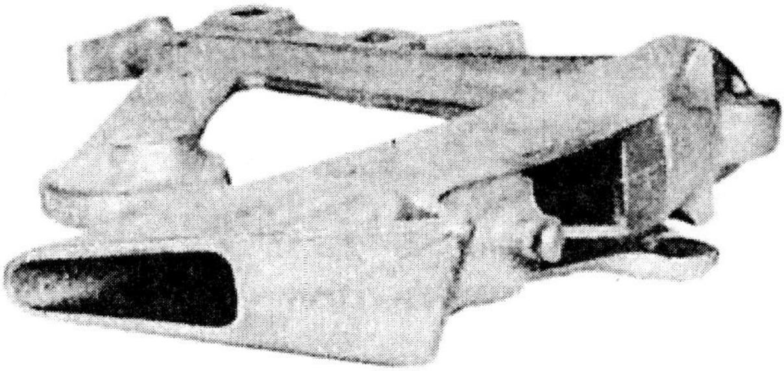

Fig. 1-4. The protruding section of the bar on this casting tended to break off during cleaning, milling or shakeout.

10. Pouring Practice

- Pouring temperature was insufficient to promote proper collapsibility of sand and cores

11. Miscellaneous

- Shakeout too hot or rough
- General carelessness in handling at shakeout
- Improper packing of castings in tumbling barrels or barrel blast
- Careless and rough handling in loading or tumbling with light thin-section castings
- Careless and rough handling in loading or unloading of cleaning equipment
- Loading or tumbling light and heavy castings in the same batch.
- Improper piling or stacking of castings
- Banding too tight on pallets
- Any mechanical handling that drops castings from a distance into containers

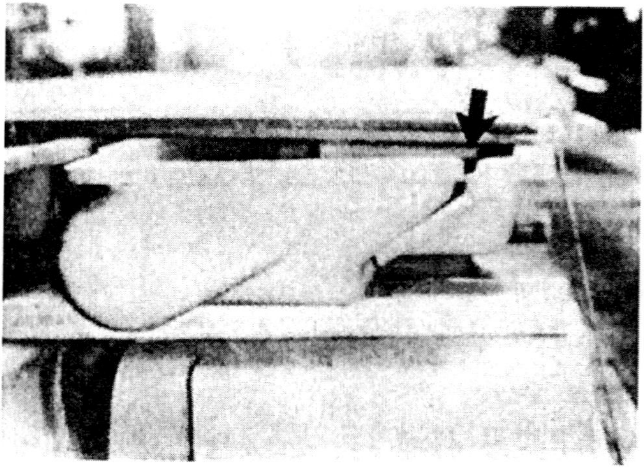

Fig. 1-5. Excessively tight banding of pallets resulted in broken castings.

CHAPTER 2

Crushes, Pushups and Clamp-Offs

Description

Crushes, pushups and clamp-offs are indentations or flaws in the casting surface. These defects are caused by disruption of the mold surface due to external or internal force or weight problems with the core system. The major causes of these defects are carelessness, particularly related to flask equipment, handling devices and molding practice.

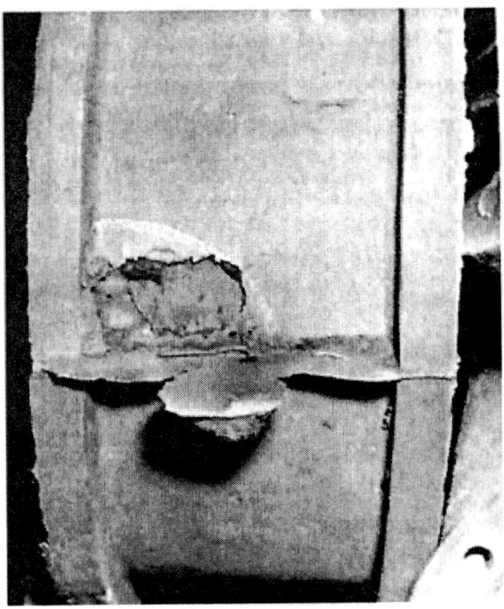

Fig. 2-1. Careless handling of the mold during assembly resulted in this crush on the cope section.

Causes

1. Casting and Pattern Design

- Insufficient draft on the pattern or the corebox
- Coreprint too small to support the core size/weight
- Lack of crushing strips on the pattern tooling

2. Pattern Equipment

- Worn patterns and coreboxes. A worn pattern results in the coreprint being too small for the normal core size. Conversely, a worn corebox results in the core being too large for a normal coreprint.
- Pattern tooling incorrectly mounted
- Misaligned cope and drag patterns or plates
- Worn alignment pins and bushings in either the pattern or corebox
- Coreprints improperly marked
- Warped or untrue pattern plate (cope and drag)
- Excessive flexibility in the matchplate
- Flask landing strips located too high above the parting line. This creates a condition of excessive loading at the sand-to-sand contact

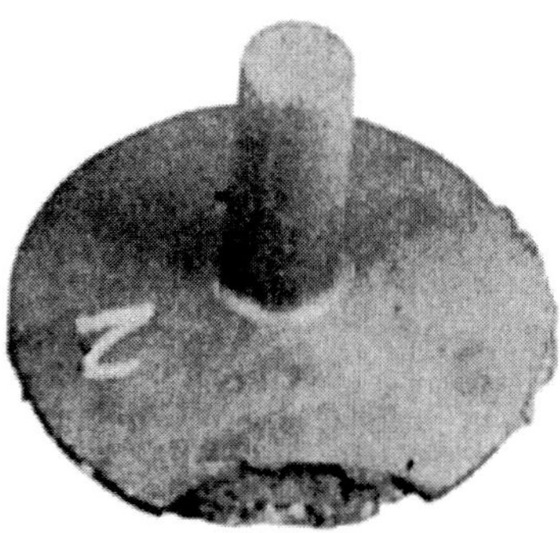

Fig. 2-2. This example of a crush was caused by poor pin alignment.

3. Flask Equipment and Handling Devices

- Misalignment of flask equipment, pins, and bushings
- Weights too heavy or uneven
- Insufficient sand bearing surface
- Warped or uneven flask joints
- Worn stripping plate on molding machine
- Improperly fitted, dirty or crooked jackets
- Worn pins and bushings
- Uneven, dirty or burned bottom boards
- Weak bottom boards
- Improperly barred cope

4. Gating and Risering

- Independently, gating and risering do not create a crush although the mounting of gate runners too close to the pattern can result in an inadequate sand bearing surface.

5. Molding Sand

- Weak sand due to poor/non-uniform compaction
- Low green tensile strength may cause the cope to sag
- Low green deformation causes the sand to crush rather than yield during flask closure
- Low dry strength in a green sand mold may fail to support normal loads

6. Core Practice

- A core too large for the coreprint will fail to seat properly and result in a crush
- A misaligned core-mold assembly will cause incorrect contact when the flask is closed
- A core sagged out of shape may be the result of:
  - Rough handling of the core while green
  - A soft ram or poorly blown core
  - Low green strength
  - Excessive water in the core mix or core wash
  - Warped driers
  - Low warm strength (i.e., excessive solvent in the core wash)
  - Improper curing or drying
  - Reinforcement omitted or insufficient (i.e., rods, wires, or arbors)
- Warped cores (including shell)
- An oversized core can be a result of excessive core wash

7. Molding Practice

- Careless closing of the molds
- Uneven clamping of the molds
- Faulty bedding of molds on bottom boards or plates
- Improper setting of the jackets
- Careless coresetting
- Use of the wrong chaplets or chaplets were omitted
- Setting dirty jackets (i.e., sand or metal buildup)
- Stem chaplets improperly wedged or seated
- Unshaved coreprints and joints
- Failure to wedge flask joints
- Careless mold handling during carry-out or on the conveyor
- Dropping or uneven weight placement

Fig. 2-3. This pushup was caused by a sand lump located on the bottom.

Fig. 2-4. A crush on a camshaft gear blank; the finished casting is located on the left.

8. Metal Composition

- None

9. Melting Practice

- None

10. Pouring Practice

- A ladle or heavy object was set on the mold surface

11. Miscellaneous

- Rough handling during any part of the molding process can break or crush a mold

Fig. 2-5. Uneven clamping of the small mold crushed one side of the mold face.

CHAPTER 3

Cuts or Washes

Description

Cuts or washes are rough spots and areas of excess metal caused by erosion of the mold or core surface by metal flow. This definition differentiates between a cut and an erosion scab, in spite of the fact that the two defects are often similar or identical in general appearance. It is necessary to make such a distinction because the cure for the two defects may be completely different.

Causes

1. Casting and Pattern Design

- A design that promotes excessive metal flow over any given area of the mold surface
- A design that includes unavoidable nozzle effects within the casting

2. Pattern Equipment

- Pattern layouts that prevent correct gating. Equalized flow is essential if excessive localized heating of the sand in the mold or core is to be avoided.

3. Flask Equipment and Handling Devices

- Flask is too small to permit proper gating
- Flask bars are located too close to pattern

4. Gating and Risering

- A gate that forces the metal to impinge on a mold or core surface (i.e., gating into a thin casting wall)
- A downsprue that does not provide a pool or well of metal ahead of the gating system
- Too much metal flowing across any mold surface will eventually redissolve the metal skin that is formed during the early stages of pouring
- Excessive metal velocity across a given mold surface
- Unequal distribution of metal through the gating system

Fig. 3-1. A drag view of cuts and washes caused by low strength sand.

5. Molding Sand

- Inadequate hot strength
- Insufficient clay content
- Inadequate water content to properly activate the clay
- Improper mixing
- Excessive use of certain additives, such as cellulose
- Hot molding sand
- Low hot strength self-curing binders

Fig. 3-2. Inclusions caused by cuts and washes on the drag side.

6. Core Practice

- Soft core surface
- Overcuring or undercuring of cores
- Insufficient amount of binder or water
- Improper mixing
- Improper application and control of core wash
- Vents or reinforcements are located too close to the surface
- Poorly patched cores or filled hook holes
- Non-uniformly prepared cores
- Low-density cores
- Core was damaged during handling
- Soft cores (using self-curing binders)
  Over mixing
  Improper sand temperature
  Expired bench life
  Stripping prematurely
  Excessive variation of acid demand of sand

7. Molding Practice

- Soft or non-uniformly rammed molds. Hot strength varies greatly depending on the mold hardness or moisture
- Mold edges burned by excessive drying temperatures
- Gaggers, bars or soldiers set too close to the mold surface

8. Metal Composition

- None

9. Melting Practice

- Tap out too hot

10. Pouring Practice

- Excessive pouring temperature for mold or core materials used
- Hard pouring

11. Miscellaneous

- Excessive parting or solvent (i.e., kerosene) can materially change the hot and dry strength

Fig. 3-3. This cut was caused by low sand strength.

CHAPTER 4

Dirt, Slag and Other Inclusions

Description

Inclusions are surface or subsurface particles of sand, slag, dross, oxides or other materials embedded in the metal. These particles may originate from molds, cores, ladles, furnaces or the metal itself.

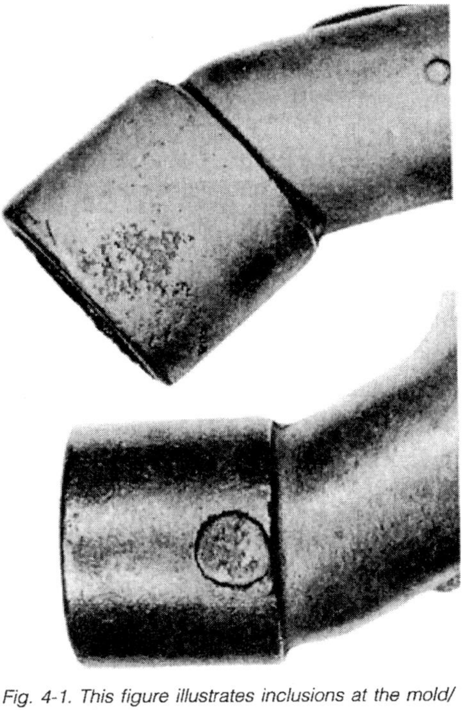

Fig. 4-1. This figure illustrates inclusions at the mold/metal interface.

Causes

1. Casting and Pattern Design

- Design factors that contribute to scabs, cuts, washes and erosion will increase the tendency to form inclusions.
- Design features that are difficult to mold, such as deep pockets or a long vertical wall, will result in areas with lower density and thus a higher susceptibility to creating an inclusion.

2. Pattern Equipment

- Incorrect pattern coating choice for the materials involved. Improperly matched coatings can cause problems such as sand adherence.
- Insufficient clearance for setting cores and mold closure
- Any factor in the handling devices that creates a rubbing or gouging effect during mold closing or coresetting
- Lack of fillets
- Insufficient draft
- Loosely mounted pattern or gating system
- Worn flask landing areas can create a crush or disrupt sand grains
- Inaccurate partings
- Entrapped pyrolysis products and glue from foam patterns.

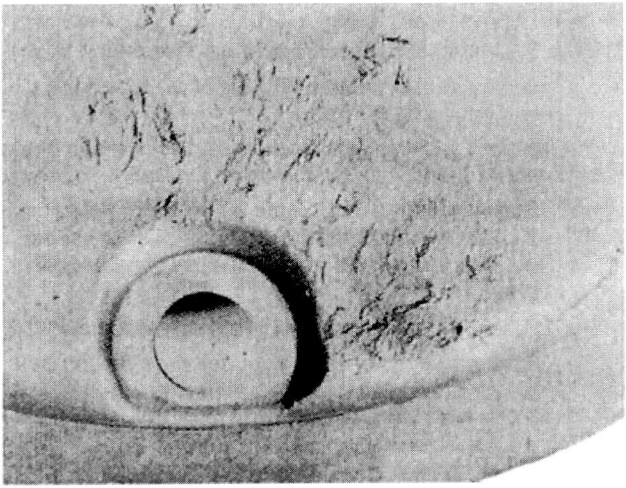

Fig. 4-2. Excessive use of liquid parting material on the pattern caused these tracks that are classified as inclusions.

3. Flask Equipment and Handling Devices

- Warped or dirty flasks
- Worn or bent pins or bushings
- Inaccurate coresetting or core assembly
- Dirty pins and bushings

4. Gating and Risering

- Factors causing scabs, cuts, and washes
- Gating design caused turbulent flow resulting in gas entrapment
- Gating systems that create slow pouring can cause scabs particularly on the cope surface
- Gating systems that create a non-filled runner should be avoided since the cope of such runners will eventually scab and drop sand into the gating system
- Gating that promotes high-velocity metal flow
- Exothermic materials that permit the reaction by-products to enter the mold or casting

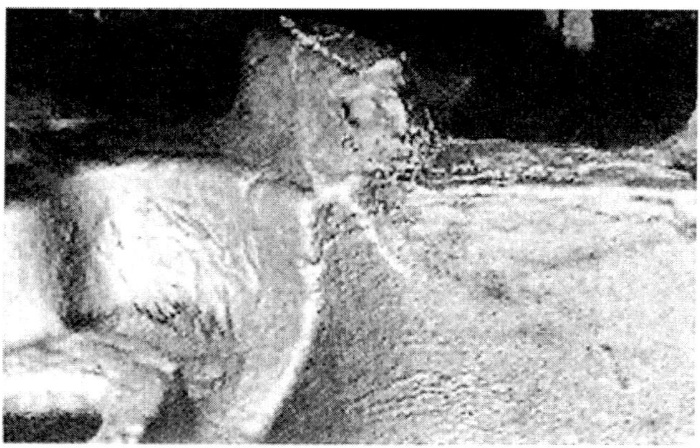

Fig. 4-3. An ineffective slag trap permitted ladle slag to locate at the top of this boss.

Fig. 4-4. This defect was caused by a contaminant in the molding sand.

5. Molding Sand

- Sands with low mechanical properties, such as low green, dry and hot compressive strength
- Mold metal gas reaction
- Coatings too heavy, not dried, or unsuitable for the application

6. Core Practice

- Weak cores
- Excessive core gas-metal reaction
- Improperly cleaned cores
- Premature collapse of cores in molds and gating system
- Core wash too heavy, not dried, or unsuitable for the application
- Broken cores

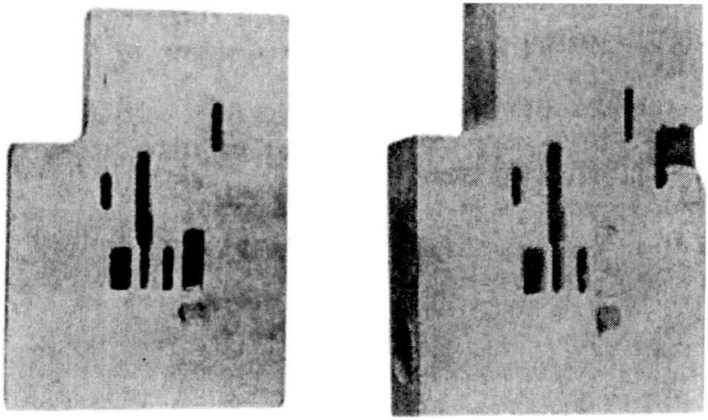

Fig. 4-5. This casting defect is a result of a broken core.

7. Molding Practice

- Any molding factor which contributes to displaced or loose sand in the mold cavity
- Inaccurate or careless molding
- Excessive use of liquid parting
- Loose sand from cores or molds

Fig. 4-6. This defect is a result of coarse seacoal in the molding sand.

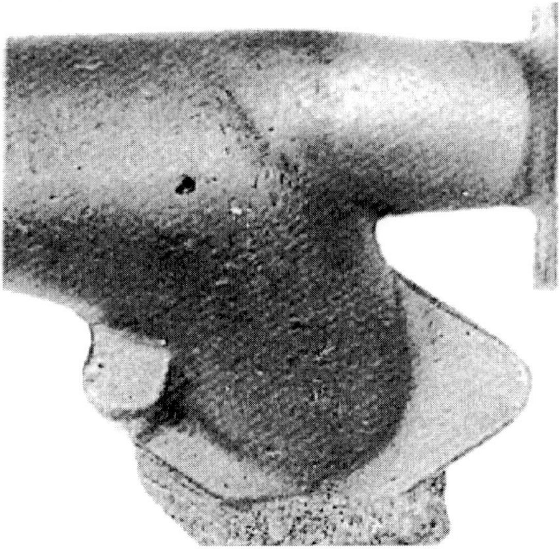

Fig. 4-7. This broken mold edge was caused by low strength sand, which permitted sand grains to wash into other casting sections.

Fig. 4-8. Loose sand remained in the mold causing these sand inclusions on the drag face.

8. Metal Composition

- None

9. Melting Practice

- Undissolved alloy addition, inoculant, or deoxidizer
- Dross resulting from the alloy addition or inadequate fluxing
- Excessively low melting temperatures may result in the inability of the slag or dross to separate
- Insufficient slag cover during melting to maintain metal cleanliness
- Thin watery slag
- Refractory-metal reaction can easily release impurities into the metal
- Excessive fluxes or deoxidizers
- Improper deoxidation or inoculation practice

Note: Slag is always generated during melting. The key is to prevent the slag from traveling downstream and entering the mold cavity. These melting practices issues undoubtedly generate excessive slag and should be avoided.

10. Pouring Practice

- A wet or boiling lip on a ladle can create, or interfere with proper separation of slag
- Careless skimming of the ladle can result in excess slag buildup in the ladle thus allowing the slag to be carried into the mold
- Low pouring temperature
- Intermittent or slow pouring
- Dirty ladles
- Improper ladle lining material that spalls or reacts with the metal

11. Miscellaneous

- Careless maintenance, handling, and general housekeeping that allow foreign material to enter and remain in the mold cavities

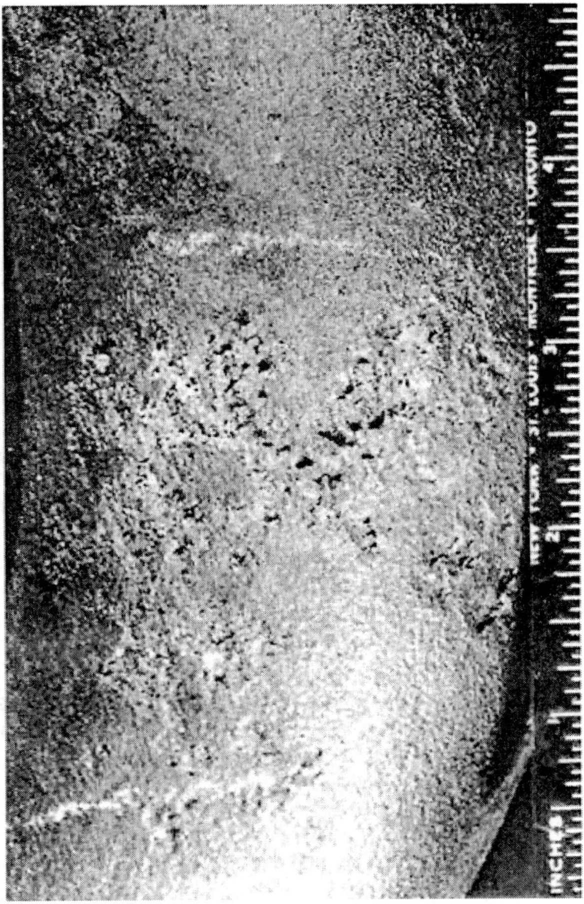

Fig. 4-9. Metallic oxides from the ladle entered the mold causing this defect. Improved ladle design or proper gating can prevent this defect from occurring.

Fig. 4-10. A poorly designed pouring basin prevented proper choke and permitted slag to enter mold.

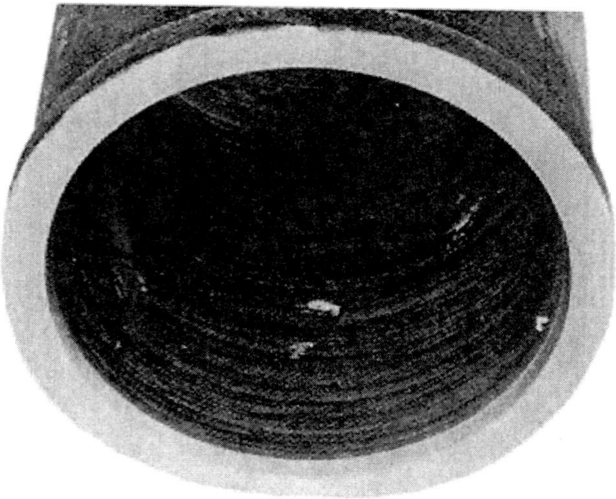

Fig. 4-11. A section of a centrifugal casting shows slag inclusions originating from refractory lining material.

Fig. 4-12. Intermittent pouring trapped slag in the metal stream and permitted it to enter the mold.

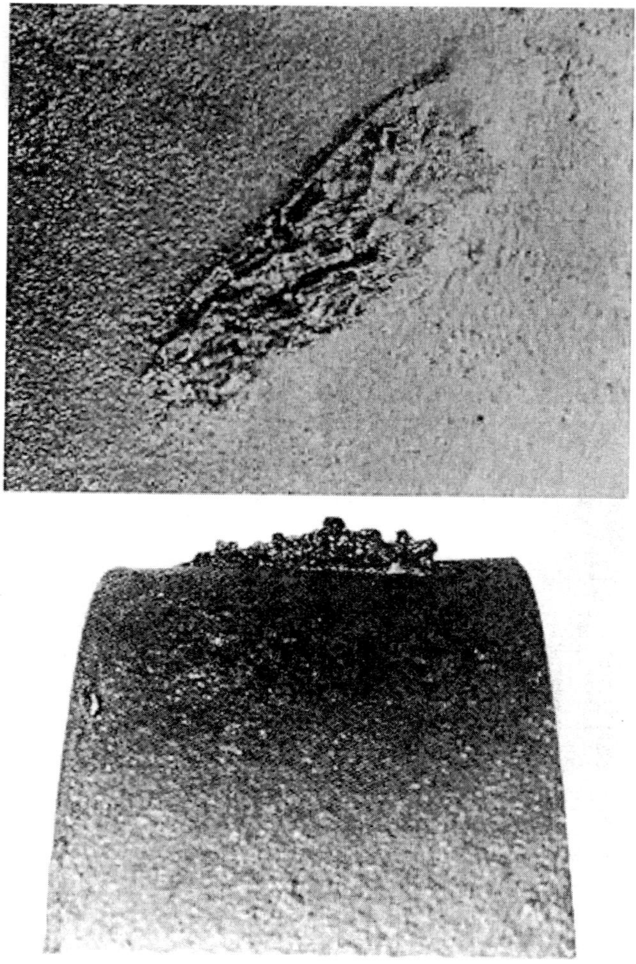

Fig. 4-13. Two views of an inclusion caused by a rag mistakenly left in the mold during casting.

Fig. 4-14. Two views of core wires in casting cavities.

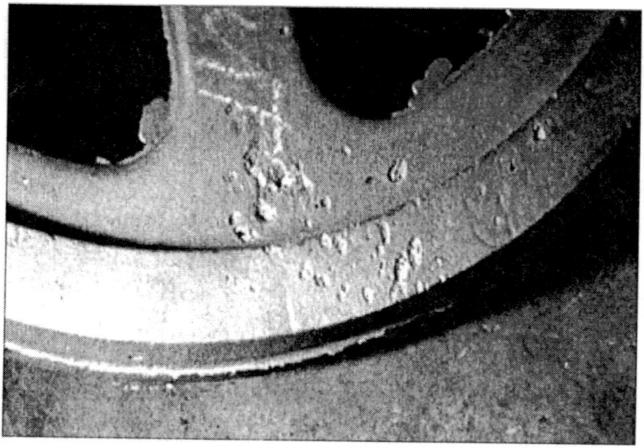

Fig. 4-15. A poorly skimmed ladle permitted slag to enter the mold cavity.

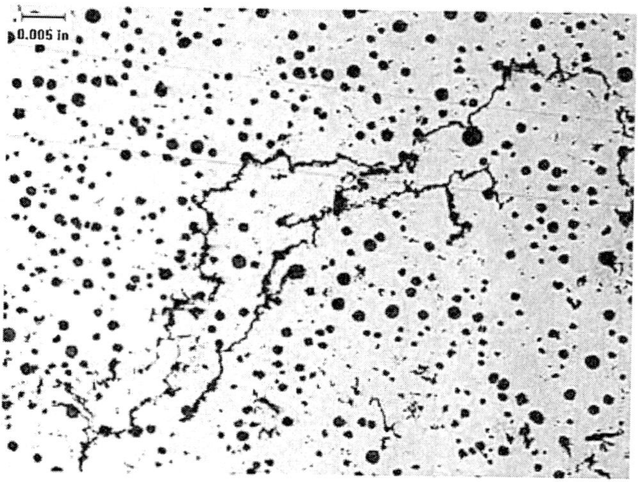

Fig. 4-16. An oxide film inclusion stringer from turbulent flow in the gating system is shown.

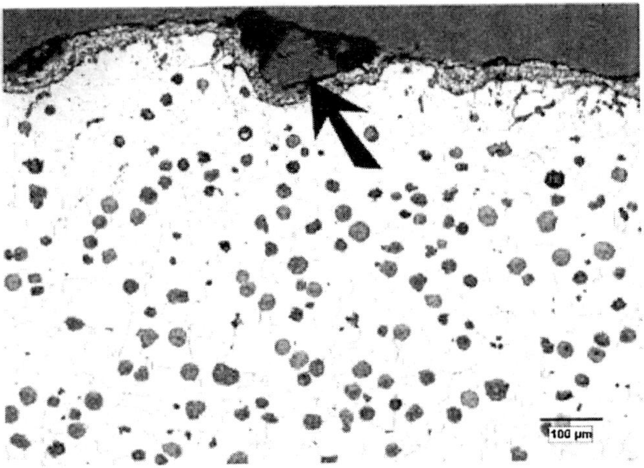

Fig. 4-17. A sand inclusion that reduced machinability.

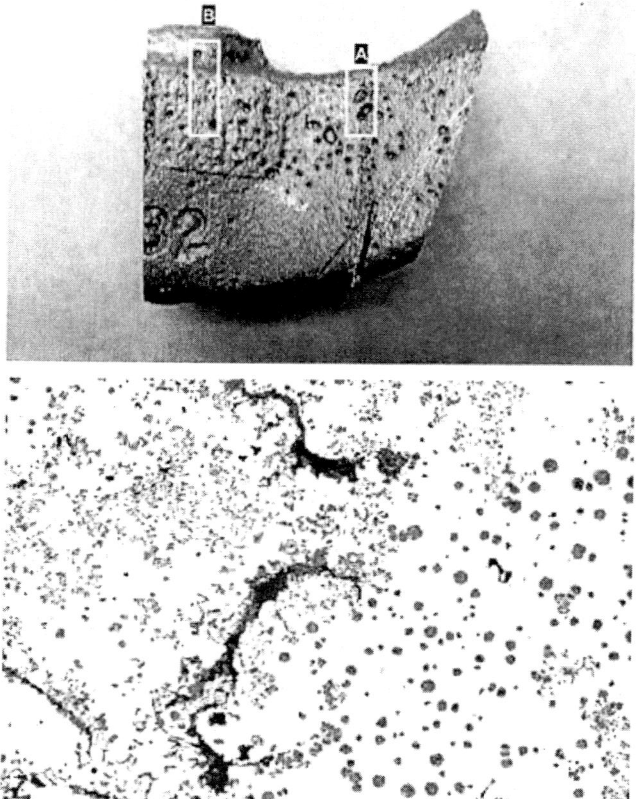

Fig. 4-18. Pinholes and dross and their corresponding cross section in the casting.

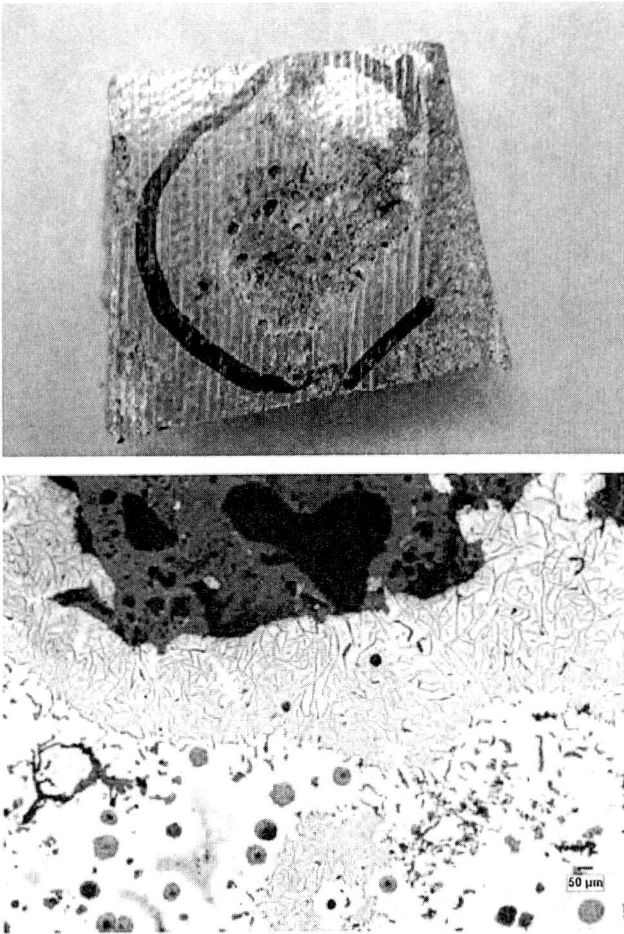

Fig. 4-19. An example of slag entrapment and the corresponding cross section in the casting.

CHAPTER 5

Drops

Description

A drop is a casting defect due to the loss of a portion of sand from the cope or other overhanging section. In appearance, the defect resembles a sticker.

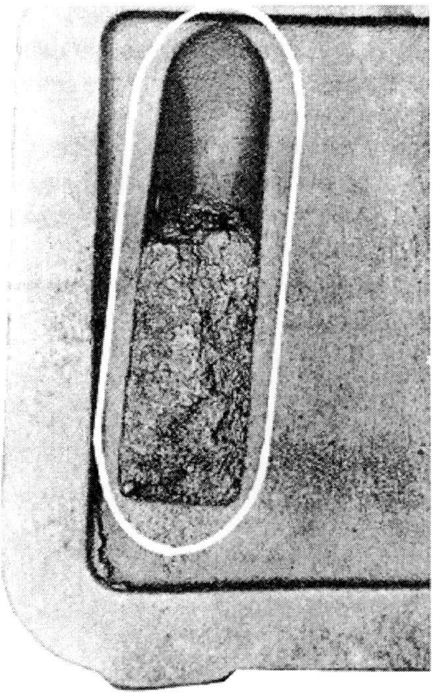

Fig. 5-1. A typical result of drop in one section of a green sand pocket.

Causes

1. Casting and Pattern Design

- Deep pockets or inadequate draft
- Protruding or overhanging sections

2. Pattern Equipment

- Insufficient draft or backdraft
- Improperly designed parting creates narrow pockets

3. Flask Equipment and Handling Devices

- Excessive number of patterns for the flask size
- Insufficient cope height
- Weak, loose, or springy flasks and bars
- Misaligned, bent or sprung bands can result in the lack of rigidity
- Not enough bars or improperly designed bars
- Rough handling due to faulty equipment
- Improperly fitted, dirty or crooked jackets
- Weights too heavy or uneven
- Flask bars located too close to risers or sprues
- Careless handling of weights or jackets

Fig. 5-2. Backdraft on the cracking strip weakened a sand pocket and resulted in a drop.

4. Gating and Risering

- Any placement of runners and risers that results in narrow sand pockets.

5. Molding Sand

- Inadequate green properties of the molding sand
- Inadequate mold properties related to sand moldability
- Low mold hardness or insufficient mold strength
- Low dry strength (dry sand molding)
- Poorly prepared sand
- Molding sand too wet

6. Core Practice

- A weak or inadequately secured core or core assembly
- Core failure in a centrifugal casting appears as a drop

7. Molding Practice

- Nonuniformly rammed or soft rammed molds
- Gaggers, nails, soldiers, bars, etc. located too close to the mold cavity
- Insufficient mold reinforcement
- Careless coresetting may fracture a portion of the mold surface
- Improper closing of the mold
- Rough handling
- Separation of molded layers of sand (i.e., separation between a facing and backing and insufficient venting)

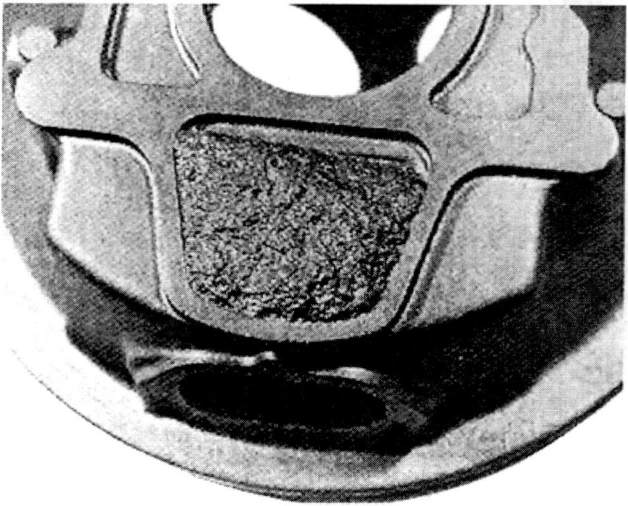

Fig. 5-3. Inadequate draft in the pocket contributed to this drop.

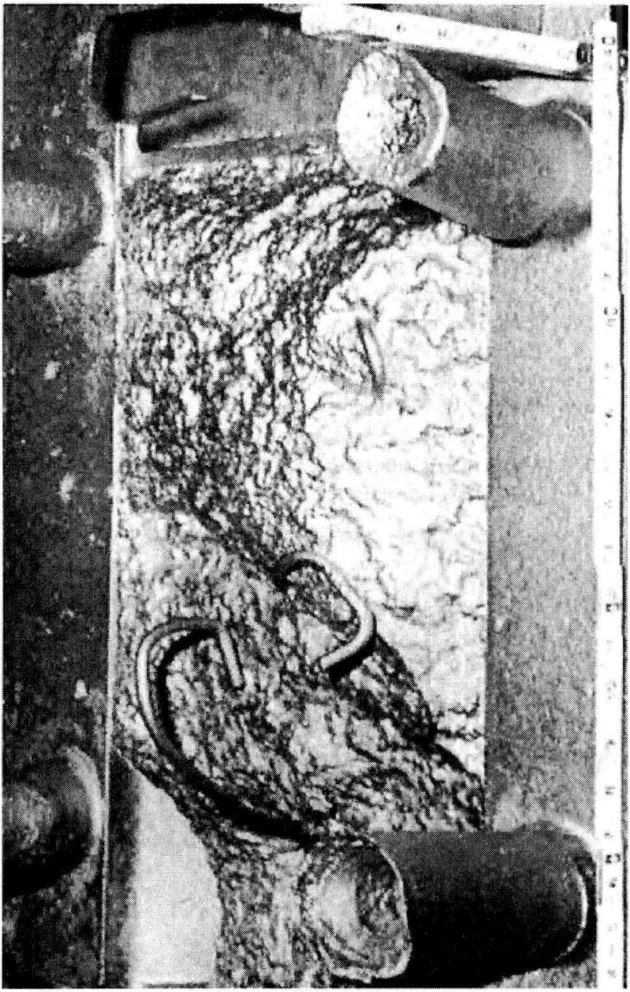

Fig. 5-4. A non-uniform ram was the major cause of drop in this large mold section.

8. Metal Composition

- None

9. Melting Practice

- None

10. Pouring Practice

- Bumping molds with the ladle

11. Miscellaneous

- Foreign material in sand
- Mold explosions that occur when the gas ignites
- Vibration caused by heavy jolt machines or other equipment

Fig. 5-5. Inadequate green compressive strength in molding sand resulted in this defect.

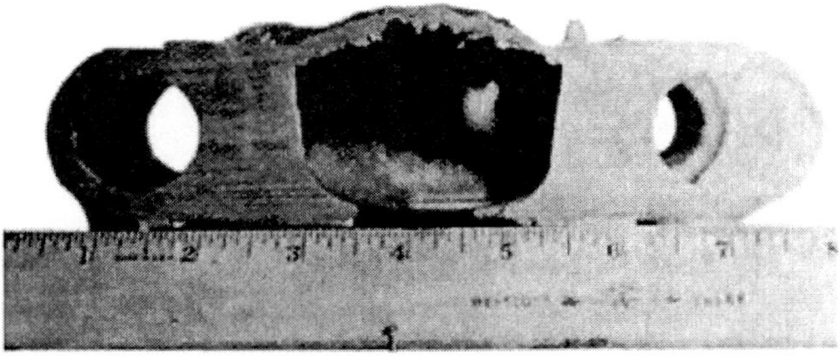

Fig. 5-6. Low green shear strength of the molding sand permitted the pocket to shear under the drop and float.

Fig. 5-7. The core was broken off during setting and dropped into the drag. During pouring the sand section floated to the cope.

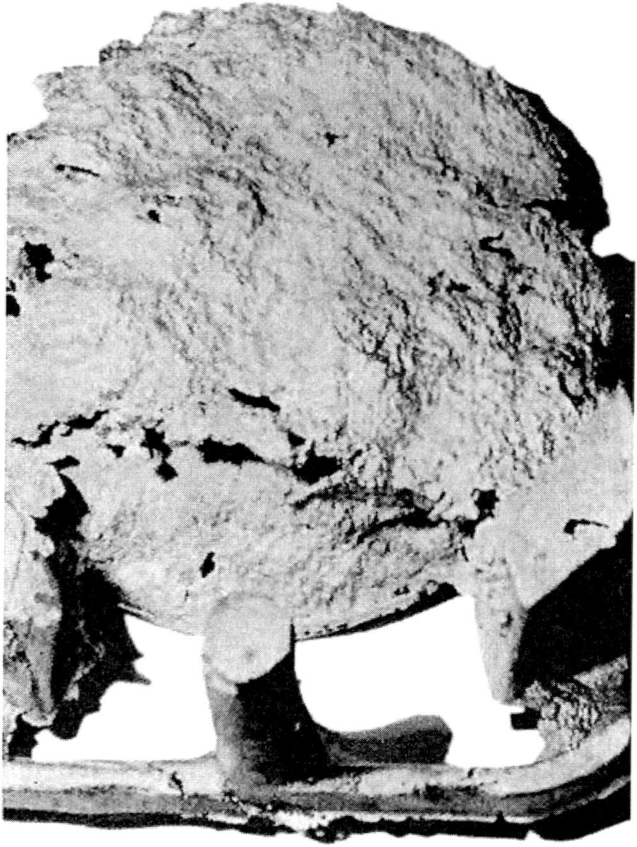

Fig. 5-8. A weight was carelessly dropped on the mold causing this defect.

CHAPTER 6

Erosion Scabs

Description

An erosion scab defect usually occurs in the drag, where loosened sand has been eroded away by the motion of the metal, and has left a solid junction between the casting and the defect. The defect can be removed usually leaving a solid surface. The erosion scab may result in sand holes or sand inclusions in some other area of the casting.

Causes

1. Casting and Pattern Design

- A design which necessitates gating through thin sections

2. Pattern Equipment

- Insufficient fillets on gates and abrupt section changes
- Parting the pattern to necessitate gating through thin sections

3. Flask Equipment and Handling Devices

- Bars improperly placed, preventing uniform ramming
- Bars, soldiers, and gaggers coated with wet clay and located too close to the mold surface
- Handling devices in an undersized flask cause the vertical pattern wall to be located too close to the flask or bar
- Plugged flask vents

Fig. 6-1. This typical scab located on the cope surface is caused by excessive slicking. Slicking is the process of smoothing/mending the mold surface with a hand tool after the pattern is withdrawn.

Fig. 6-2. A scab caused by metal impingement on the sand surface (gates are located on the inside of the two bosses).

4. Gating and Risering

- Gating that creates an interrupted metal flow
- Metal impinging on and disrupting the sand surface
- Gating which creates localized overheating of the sand

Note: There is a fine line of distinction between an erosion scab and a cut or wash. As far as metal flow alone is concerned, the two defects have the same cause. In the case of erosion scab there is not enough impingement to actually cut, in the absence of expansion, but is enough when expansion stresses exceed the strength of the sand.

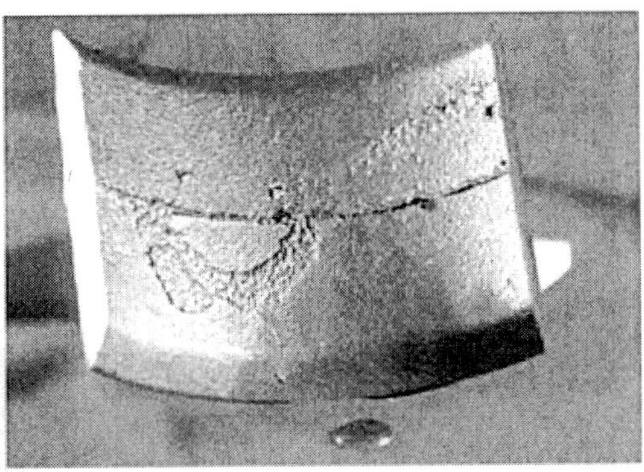

Fig. 6-3. Two views of a casting showing that the erosion of sand can occur even in strong molds. For example, an undersized ingate can produce a high-velocity metal stream and cause erosion.

5. Molding Sand

- High moisture content
- Excessive volatile material
- Insufficient carbonaceous (or cellulose) material for expansion control
- Insufficient hot deformation
- Improper conditioning of the sand, such as clay or moisture not uniformly distributed
- Under tempered sand

6. Core Practice

- Excessive moisture in or on the core
- A core blow may cause an erosion scab on the adjacent mold surface
- Insufficient hot deformation
- Core wash not properly dried
- Insufficient penetration of the wash
- Undercured cores
- Excessive core wash

Fig. 6-4. Excessive pouring temperatures caused erosion scab on the cope surface of this casting.

7. Molding Practice

- Non-uniform ramming
- Gaggers located too close to pattern
- Excessive clay on gaggers
- Excessive slicking
- Mold wash not properly dried
- Insufficient penetration of wash
- Excessive wash

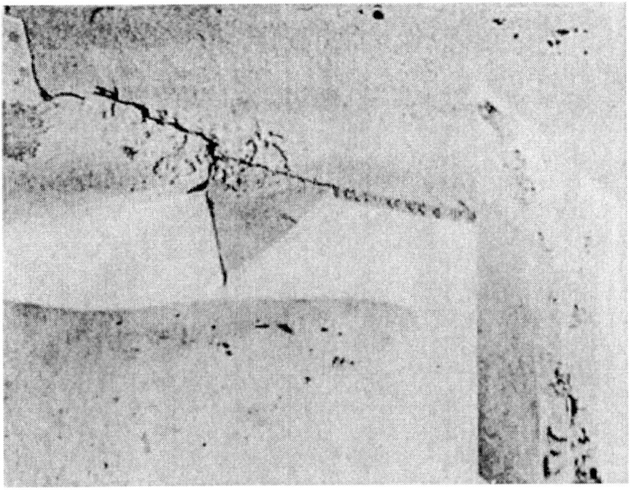

Fig. 6-5. This erosion scab was caused by wet sand and aggravated by hard ramming.

8. Metal Composition

- None

9. Melting Practice

- None

10. Pouring Practice

- Excessive pouring temperatures
- Interrupted or slow pouring

11. Miscellaneous

- Any foreign body located close to the mold surface

CHAPTER 7

Expansion Defects

Description

Although the following four stages of development present different appearances, the cause for the expansion defect is the same. As the intensity of the cause increases, the severity of the defect graduates from Stage One through the various stages to Stage Four.

Stage One: Rattails which are irregular lines caused by low temperature expansion of the mold surface resulting in a fault in the mold surface.

Stage Two: Buckles which are v-shaped indentations which may occur independently or under an expansion scab.

Stage Three: Expansion Scabs which are rough layers of metal connected to the main body of the casting by a vein of metal.

Stage Four: Cope Spalls (or Pull Downs) which are indentations in the cope surface of the casting.

Depending on the time of formation these defects may have the appearance of a buckle, rattail, shrink, or blow. Shrinkage cavities are voids in castings that appear as a result of solidification shrinkage and the progressive freezing of metal towards the center. Blows are gas defects from the trapping of gas in molten or partially molten metal, causing voids in the casting. Coating scabs (also called blacking/blackening scabs) are a special form of a scab in which the defect is related to the coating rather than to the sand. Since sand surfaces go through the process of expansion and contraction during the pouring and solidification of the casting, it is a common practice to assume that the sand is the principal cause. This is a mistake, as these defects are closely related to all practices.

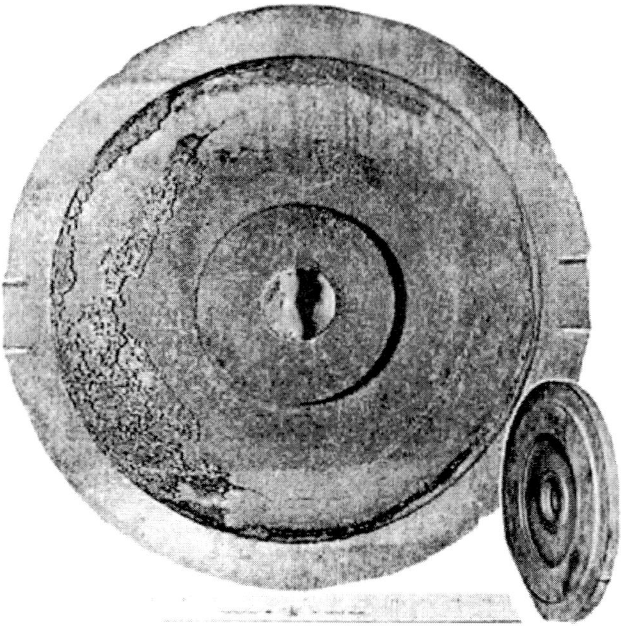

Fig. 7-1. A typical expansion defect on the cope surface.

Fig. 7-2. A typical rattail resulting from an interrupted metal flow.

Causes

1. Casting and Pattern Design

- Large uninterrupted flat surfaces
- Inadequate radii
- Large, smooth uninterrupted concave or convex surfaces

2. Pattern Equipment

- Inadequate fillets

3. Flask Equipment and Handling Devices

- Bars or flask located too close to pattern surface
- Handling device which prevents proper pouring speed
- Flask too shallow

4. Gating and Risering

- Pouring speed is hindered by incorrect gating design
- Gating that causes interrupted metal flow
- Poor design of the gating system causes nonuniform heating of the mold surface

5. Molding Sand

- Excessive moisture content for the materials involved
- Clay content too low for the moisture involved or the mold hardness
- Carbonaceous material level too low
- Low hot deformation
- Incorrect sand grain distribution
- Insufficient cushioning material
- Segregation of materials by improper mixing
- Low dry and hot compressive strength
- Insufficient new sand additions

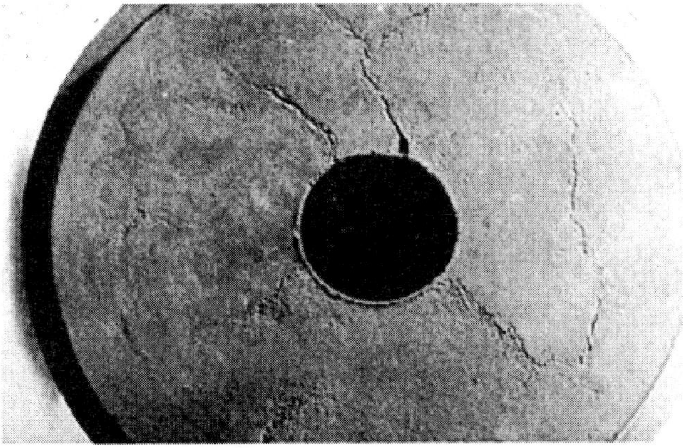

Fig. 7-3. These irregular-shaped lines are rattails resulting from thermal deformation of the sand mold during filling.

Fig. 7-4. This buckle with a scab (before cleaning) was caused by inadequate carbonaceous or cushioning material.

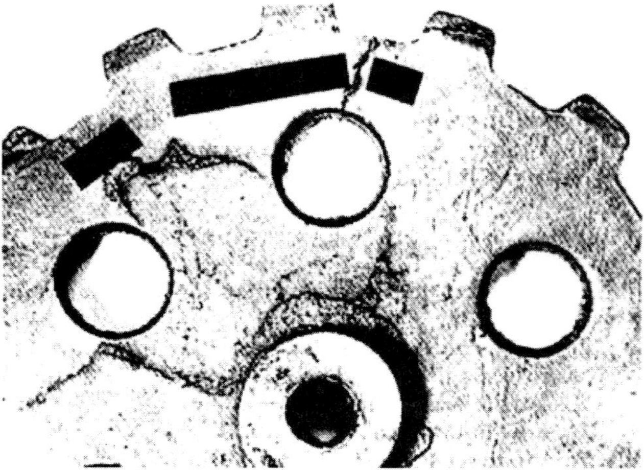

Fig. 7-5. This scab (with buckle on opposite side) is related to inadequate hot deformation.

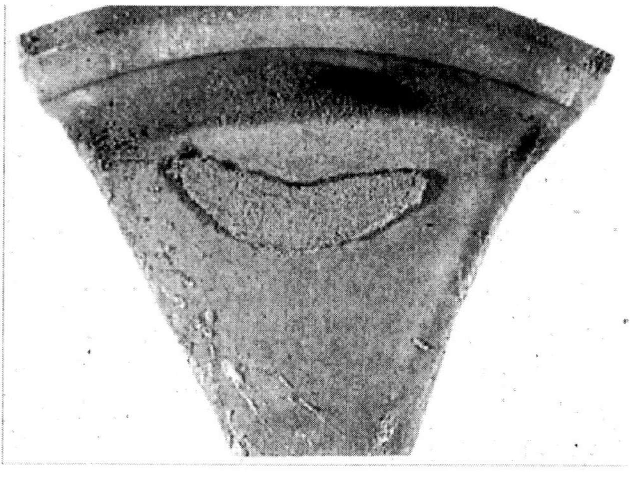

Fig. 7-6. An expansion scab defect is shown in this sectioned casting.

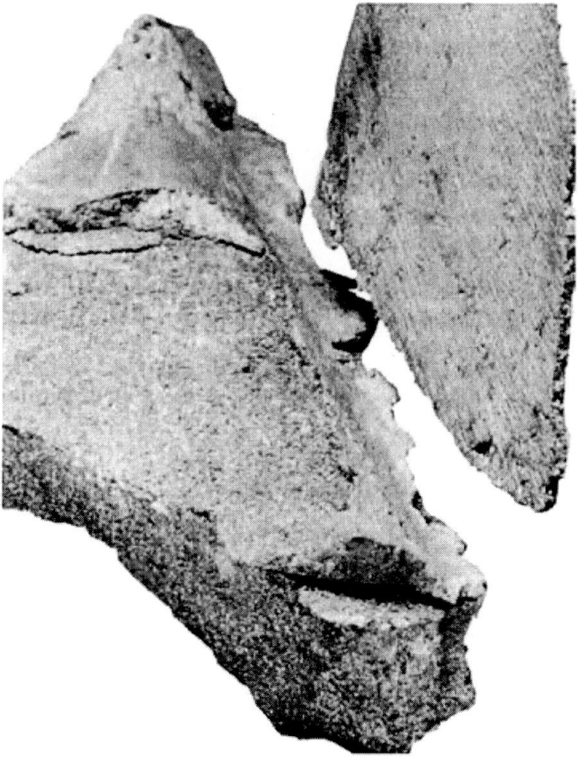

Fig. 7-7. A coating scab (also called blacking/blackening scab) caused by inadequate wash penetration.

6. Core Practice

- Low hot deformation
- Green sand or green top cores
- Nonuniform shell thickness
- Rods or arbors located too close to the core surface
- A nonuniform ram
- Insufficient penetration of wash
- Improper type of wash
- Excessive wash
- Insufficient drying of wash
- Poor sand grain distribution

7. Molding Practice

- Nonuniform or hard ramming
- Insufficient penetration of wash
- Excessive wash
- Insufficient drying of mold or wash
- Gaggers, soldiers, and bars located too close to mold surface
- Excessive troweling and slicking
- Downsprue and risers located too close to bars and flask

Fig. 7-8. Expansion scab in a dry sand mold after removal from the casting. This defect was caused by excessive clay wash on a gagger located close to the casting surface.

Fig. 7-9. Uneven ramming of a drag mold allowed the pattern plate to bend and prestress the mold surface (defect corrected by hard ramming the drag).

8. Metal Composition

- Metals vary in their tendency toward expansion defects as they vary in temperature and pouring rate

9. Melting Practice

- Factors that control fluidity

10. Pouring Practice

- Pouring too slow
- Pouring too hot

11. Miscellaneous

- Foreign material between the centrifugal mold and sand, which appears similar to an expansion type defect

CHAPTER 8

Gas Defects

Description

Gas holes in castings are cavities, spherical, flattened or elongated in shape. They are related to localized gas (including entrapped air) pressure that exceeds metal pressure in any locality during solidification of the metal or are the result of a reaction between the metal and foreign material in the metal or the mold. One of the greatest problems with core blows (captured gas defects from outgassing core material) is the ease with which they may be confused with mold blows (captured gas defects from entrapped air or gases generated from mold materials). Since gas travels upwards in the mold it is quite possible for a gas bubble to form from a core and then travel toward the cope surface.

Pinholes, blisters, and body scars as well as certain types of porosity are variations of gas holes. In gas defects, there are several possible sources, among which are cores, molding sand and metal. The volume of gas may be kept unchanged but offset by providing easier escape or by providing more metal pressure, forcing gas through existing vents. Reducing the presence of slag and slag forming conditions can also prevent the formation of gas defects (*see Chapter 4, Dirt, Slag & Other Inclusions*).

Fig. 8-1. Pinholes caused by excessive moisture in the molding sand combined with low permeability and poor grain distribution (6 screen sand).

Causes

1. Casting and Pattern Design

- Insufficient print or outlet provided for venting
- Design that prevents adequate metal pressure
- Insufficient vents at the parting line

2. Pattern Equipment

- Density of the foam pattern too low
- Foam pattern coating permeability too high
- Foam pattern coating thickness too thin
- Sharp corners on foam pattern
- Inadequate bead fusion on the foam pattern
- Excessive glue on foam pattern
- Inadequate drying of the foam pattern

Fig. 8-2. Pinholes from excessive permeability and moisture were remedied by lowering the permeability from 140 to 90-110 and reducing the moisture from 4.2% to 2.8-3.3%.

Fig. 8-3. Gas blows caused by clay balls (excess fines).

3. Flask Equipment and Handling Devices

- Bars located too close to the mold surface
- Flask wall too close to mold surface
- Cope too shallow
- Sprues and risers too close to bars or flask walls
- Insufficient venting in flasks

4. Gating and Risering

- Insufficient cope height
- Gating causing turbulence
- Gating design that causes interrupted flow of metal
- Inadequate riser height or size
- Improper sprue or gate design allowing air or mold gas to enter metal stream
- Insufficient spinning speed during centrifugal casting of pipe
- Insufficient vents and flow-offs

5. Molding Sand

- High moisture content
- Inadequate permeability for the volatile materials involved
- Foreign material in sand (i.e., shot, nails, coke, cinders, etc.)
- Clay balls
- Poorly mixed sand
- Coarse particles of gas-producing materials
- Coatings containing excessive gas-producing materials
- Excessive permeability in skin-dried molds, allowing sweat-back
- Excessive or insufficient reducing volatile materials

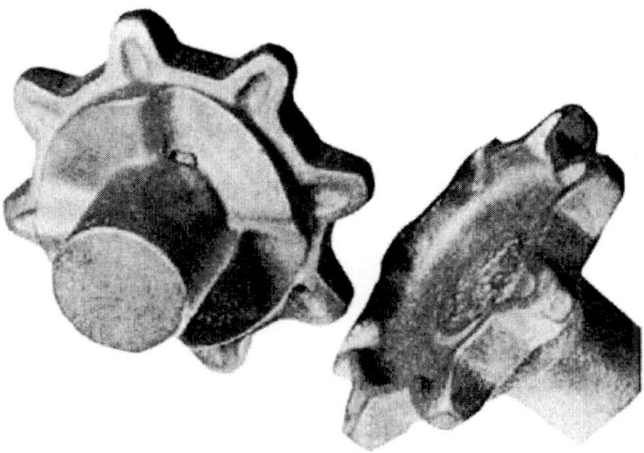

Fig. 8-4. Blow due to excessive moisture in the molding sand.

6. Core Practice

- Underbaked or undercured cores
- Excessive binder
- Inadequate permeability for conditions involved
- Inadequate venting
- Excessive or wet core paste
- Core wash too heavy, skin broken or not dried
- Excessive gas-producing material either in the core sand or wash
- Cores that absorb moisture either in storage or in the mold
- Exposed hooks, wires or hangers
- Chill or sealing materials not dried
- Hooks or hanger holes filled and not dried
- Foreign materials, such as debris, coke, coal, etc.
- Excess nitrogen content in resin binders
- Poorly mixed core sand

Fig. 8-5. Improperly conditioned core sand (lumps of cereal binder and excessive moisture are potential sources of gas and can cause blows). This defect is a gas blow from a core.

Fig. 8-6. Broken surface of core wash permitted gas to backup into heavy metal sections.

Fig. 8-7. Two views of vents plugged with iron during pouring caused a major defect on the cope surface.

Fig. 8-8. A blow in a manifold at a hot spot, caused by a hard core.

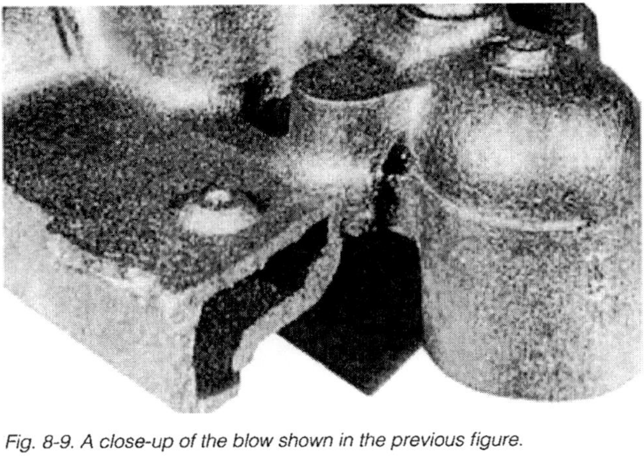

Fig. 8-9. A close-up of the blow shown in the previous figure.

7. Molding Practice

- Combination of hot and cold materials
- Improper venting in green or dry sand molds
- Insufficiently dried molds
- Hard spots caused by improper ramming for conditions involved
- Gagger, soldiers, arbors, or bars too close to pattern
- Clay wash too heavy on gaggers
- Excessive slicking and patching of green sand
- Dough roll too wet or too close to mold cavity
- Wet pouring basins and gating systems
- Insufficient sand in flask
- Excessive liquid parting (especially in pockets or depressions)
- Mold wash or coating too heavy or not dried
- Failure to lead core vents out of molds
- Vents located too close to the surface causing sweatback

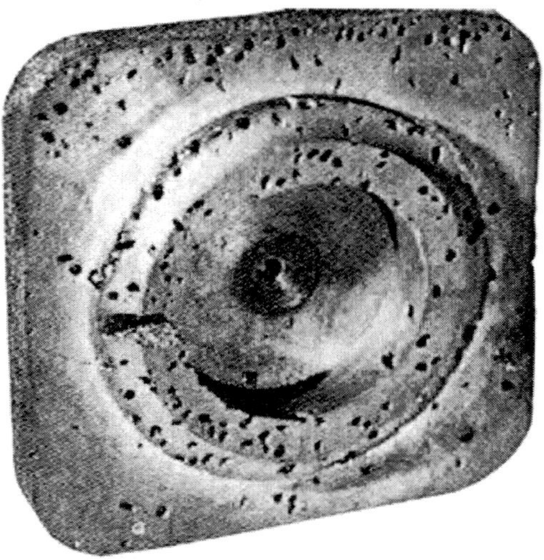

Fig. 8-10. Gas blows caused by cold iron and high moisture.

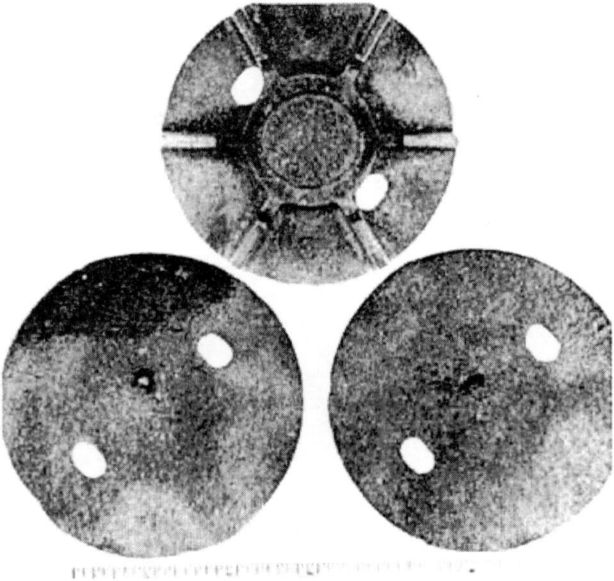

Fig. 8-11. Blow in the cope side of a pump impeller caused by wet sand and hard ramming is shown.

Fig. 8-12. This photo shows the result of aluminum in iron.

8. Metal Composition

- Gas dissolved in molten metal
- Metallic impurities, i.e., aluminum, bismuth, lead, tin, antimony and boron
- Greasy or rusty scrap
- Enameled scrap containing boron
- Incorrect balance of aluminum and silicon in steel

Fig. 8-13. This blister was caused by hard ramming and wet sand.

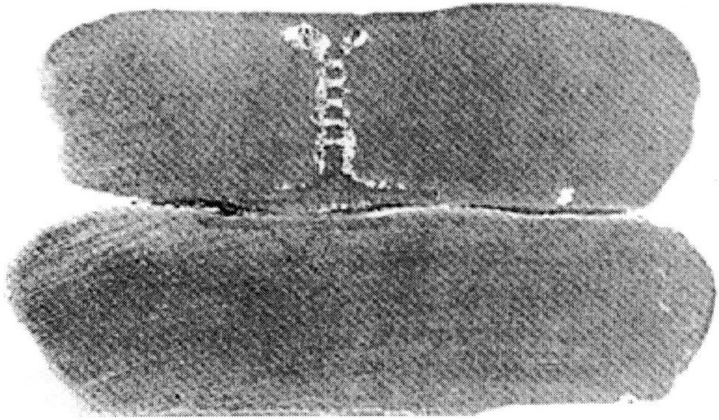

Fig. 8-14. The upper section of this casting shows a chaplet blow caused by improper coating material or a damp chaplet. The lower section shows good fusion.

Fig. 8-15. Two examples of blows caused by cold metal.

9. Melting Practice

- Steel
  - Boil too short
  - Insufficient deoxidation
  - Wet refractories
- Iron
  - Cold melted metal
  - Oxidized iron
  - Tramp elements (i.e., aluminum, lead, tin, antimony, boron, bismuth, enameled iron)

10. Pouring Practice

- Cold, damp or green crucibles or ladles
- Pouring cold metal
- Interrupted pouring
- Ladle or crucible located too high above the mold
- Slow or fast pouring
- Improperly cleaned ladles
- Heel of metal remained in the ladle
- Insufficient spinning speed in centrifugal casting
- Wet refractory in the ladle from improper curing or patching

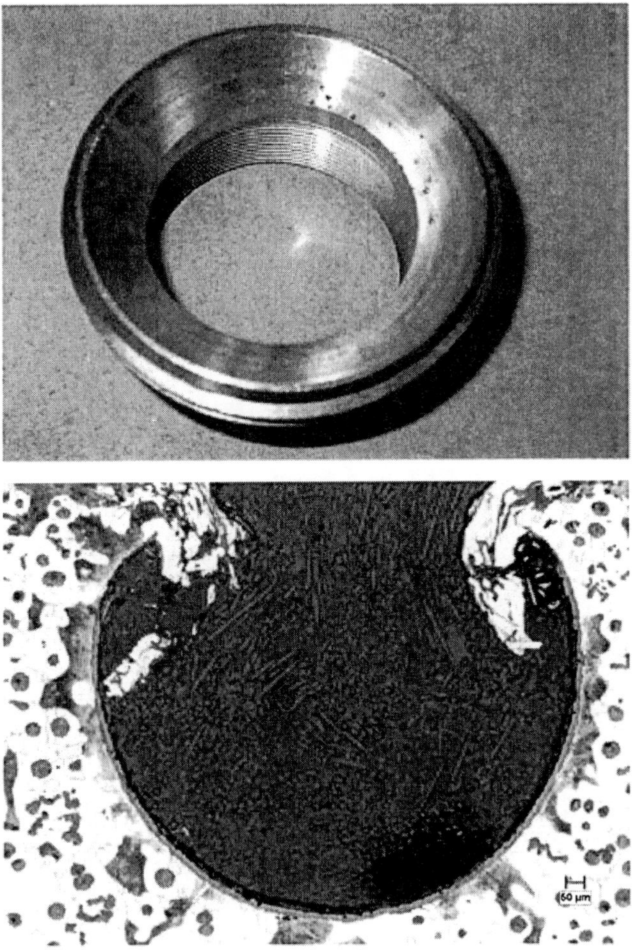

Fig. 8-16. Two views of hydrogen pinholes that were revealed after machining. The close-up view of the pinhole is spherical in shape, small in size (≤3 mm) and it is located next to the casting surface. It is typically covered with a graphite layer.

Fig. 8-17. Two views of complex gas porosity resulting from the mold (nobake, phenolic, urethane), metal and wet refractory reactions.

11. Miscellaneous

- Use of chills, chaplets or wires that are improperly coated, oxidized or damp
- Temperature difference between the metallic objects placed in the mold and the sand temperature resulting in condensation

CHAPTER 9

Gross Segregation, Carbon Flotation, Kish Graphite & Dross

Description

Segregation is the metallurgical or mechanical separation of one or more elements during solidification. Normally, this is a function of metal composition and cooling rates. Kish is free graphite that separates from molten gray iron. Carbon flotation in ductile iron is the formation of a cluster of larger size nodules of graphite in a particular location of the casting usually in hot spots towards the cope surface. Dross is a reaction product that results from magnesium treatment or turbulent flow in ductile iron castings.

Causes

1. Casting and Pattern Design

- Too great of a variance in metal section size causes segregation due to slow cooling in the larger sections if the composition is planned for the more rapid cooling sections
- A section of mold or core that promotes heat retention and could promote localized segregation
- Higher carbon percent (or carbon equivalent) for a given casting section and cooling conditions

2. Pattern Equipment

- None

3. Flask Equipment and Handling Devices

- A rough running flask (out of balance) in centrifugal castings can promote segregation

4. Gating and Risering

- Distribution of gates and risers promotes heat retention in localized area
- Localized hot spots at the gate or riser junction could slow down cooling to the point that gross segregation is found in the gate junction or riser neck
- Turbulent flow of metal changes the normal cooling rate
- Incorrect use of exothermic molding materials can cause localized overheating

5. Molding Sand

- Abnormally slow heat transfer due to the use of insulating materials or molding aggregates can change the cooling rate of a given section

6. Core Practice

- Under baked oil-bonded cores which create an exothermic reaction causing excessive heating of the core could change the cooling rate of an adjacent metal section
- Insulating gas films such as in shell cores may act to change the rate of heat transfer from the metal section into the sand

7. Molding Practice

- Introduction of any insulating material or perlite can produce localized slow cooling
- Incorrect use of exothermic molding materials changes the cooling rate of metal sections adjacent to these materials

8. Metal Composition

- In ductile iron, gray iron, and malleable, the carbon equivalent may be too high for the cooling rate involved. The eutectic point in these systems is seldom found exactly at 4.3 percent carbon equivalent
- Phosphorus level too high for the application will influence eutectic formation.
- Insufficient carbide stabilizers permit free carbon to segregate in high carbon equivalent irons
- Excessive use of graphitizers may have a direct effect on carbon flotation and kish formation in addition to their obvious influence on carbon equivalent
- Spheroidal graphite iron (ductile iron). Too high carbon equivalent. There is an added influence from the use of either excessive magnesium treatment or excessive secondary inoculation
- High sulfur can contribute to the segregation of magnesium sulfide and other segregations that are forced out of solution by the nodularizing treatment
- Aluminum plus titanium in excess of 0.10 percent produces a segregation of ternary carbide eutectics in iron
- Steel does not usually exhibit gross segregation except in high or special alloys. High sulfur or leaded steels can show segregation. Special high copper steels may also show segregation if the copper content exceeds approximately 0.6 percent.

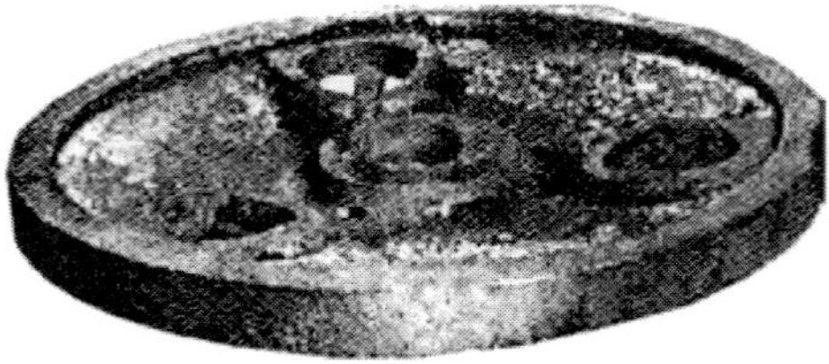

Fig. 9-1. This figure illustrates kish caused by a high carbon equivalent.

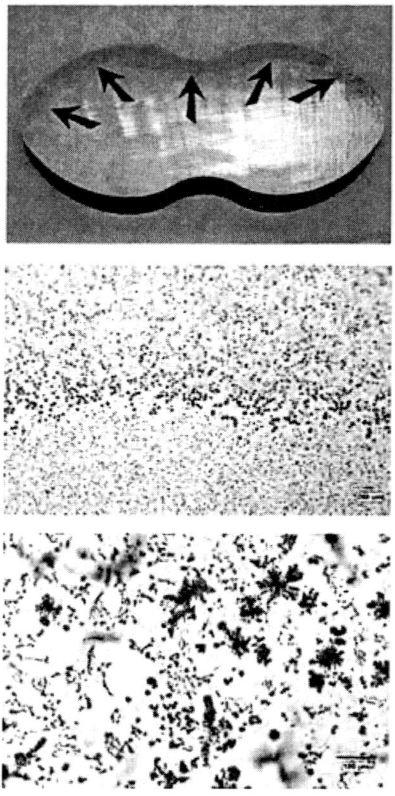

Fig. 9-2. Graphite flotation and degenerated graphite particles are shown at three magnifications.

Fig. 9-3. A view of kish graphite that resulted in graphite pullout.

9. Melting Practice

- Cupola melted gray iron and ductile iron may develop segregation as a result of a faulty melting practice which produces an unintended change in composition. Improper balance of air to coke promoting carbon pickup will produce an undesirable level of carbon equivalent.
- Melting too hot causing excessive carbon pickup at the beginning of a heat can produce an off-analysis and kish.
- Nonuniform chemical composition, due to variation in the size of charge materials, produces variations in carbon and possible segregation
- Careless charging or careless weighing of charge components results in a non-uniform chemistry
- For malleable iron, melting too hot, too fast, or with a carbonizing flame can result in excessive carbon equivalent, or may result in a direct seeding effect which subsequently starts kish accumulations
- In steel alloys, improper melting or deoxidation
- Careless scrap segregation can result in incorrect chemical analysis of the melt.

10. Pouring Practice

- Pouring below the solubility temperature is most apt to occur when alloying elements differ widely in their solubility temperature
- Cooling before pouring permits some components to come out of solution even though the original melt was adequately blended and alloyed

11. Miscellaneous

- Cooling rate too slow for the composition

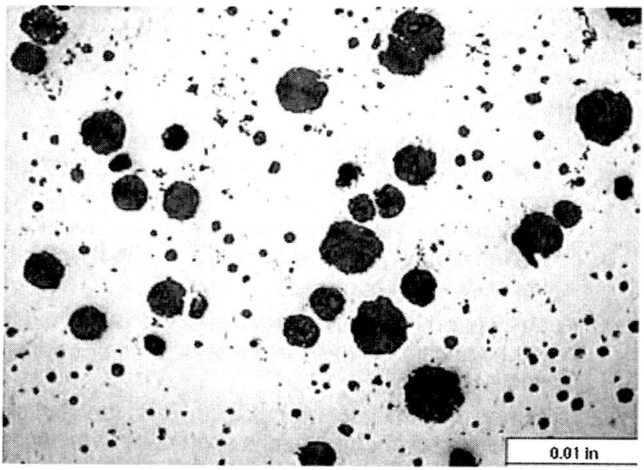

Fig. 9-4. Carbon flotation in ductile iron is shown.

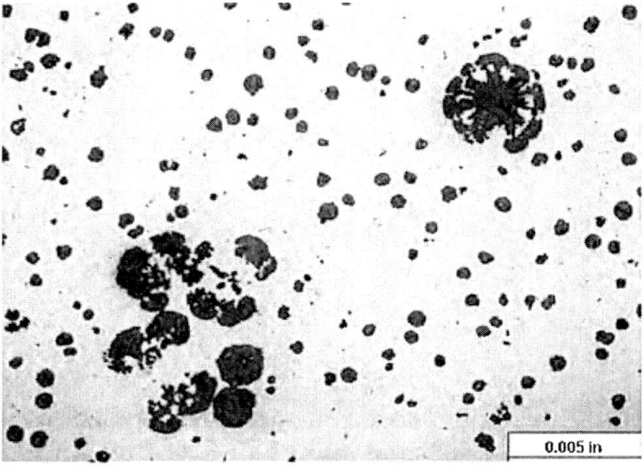

Fig. 9-5. Carbon flotation with exploded graphite is shown.

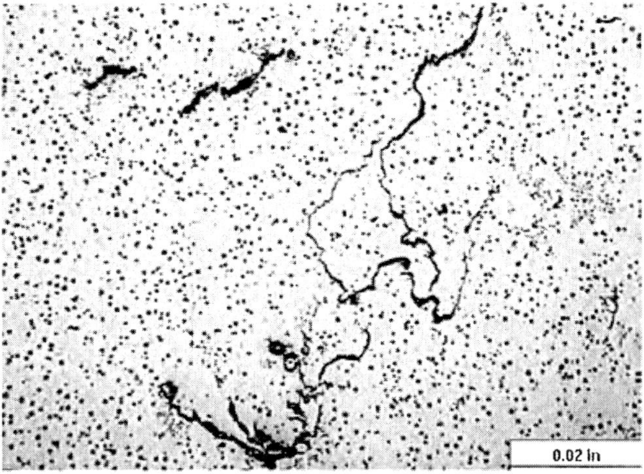

Fig. 9-6. A dross and slag slinger defect in ductile iron.

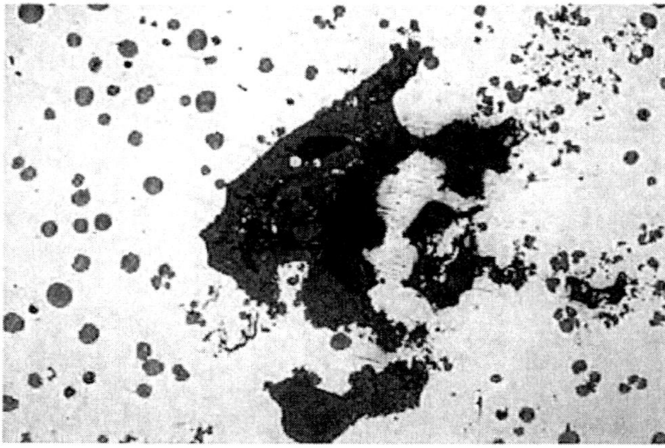

Fig. 9-7. An example of a gross inclusion is shown above.

CHAPTER 10

Non-Uniform Hardness, Hard Spots, Hard Areas & Chilled Spots

Description

Hard spots, hard areas and chilled spots in a casting are localized zones of excessive hardness. Non-uniform hardness is an excessive variation in hardness in the same areas of different mold cavities or in similarly sized areas of a casting.

Causes

1. Casting and Pattern Design

- Isolated thick and thin sections or extremely thin sections create excessive hardness from chill due to rapid cooling.
- The specified metal chemistry is not compatible with the designed casting section thickness creating excessive hardness or chill.
- Insufficient or inconsistent mold wall thickness creates excessive/inconsistent hardness or chill.

2. Pattern Equipment

- Worn or warped patterns and/or coreboxes cause thin sections or parting line flash which promotes chill carbides.
- Mismatched pattern and/or corebox equipment results in thin sections or parting line flash which creates chill carbides.

3. Flask Equipment and Handling Devices

- Mold cavity located too close to bars or flask wall/ mold wall causing non-uniform/excessive cooling rate.
- Shift created by worn flask pins and bushings or molding machine misalignment creating parting line flash (chill carbides) of thin sections (chill or excessive hardness).
- Weak or warped flasks and bottom boards

4. Gating and Risering

- Gating and/or risering that fails to provide adequate heat to isolated sections of thin sections

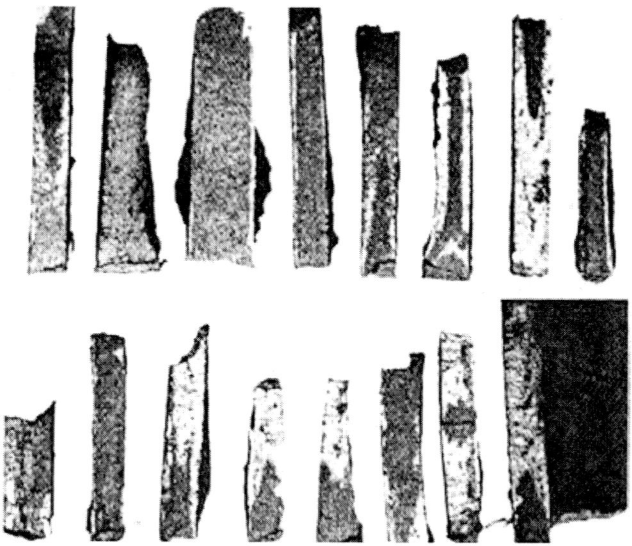

Fig. 10-1. Hard spots caused by the use of a core wash contaminated with tellurium are shown.

5. Molding Sand

- Wet spots cause excessively rapid cooling
- Tramp materials such as spill metal cause excessively rapid or non-uniform cooling

6 Core Practice

- Wires or rods located too close to the surface causing extremely rapid or non-uniform cooling rate
- Incorrect or wet coating causing extremely rapid or non-uniform cooling rate
- Warped cores causing thin sections
- Improper print size causing thin sections or parting flash that induces chill carbides

7. Molding Practice

- Careless coresetting causing thin sections or parting flash that induces chill carbides
- Arbors, rods, gaggers or bars located too close to mold cavity causing excessive or non-uniform cooling rate
- Excessive nailing
- Wash too thick and not dried causing excessive or non-uniform cooling rate
- Excessive or improper chills causing excessive or non-uniform cooling rate

8. Metal Composition

- Metal composition (carbon, silicon, carbon equivalent, pearlite stabilizing elements) not suitable for section size causing excessive hardness or chill carbides
- Tramp elements such as tellurium, bismuth, chromium, molybdenum, vanadium, boron, and niobium from the charge materials, ferroalloys, core/mold washes or spent thermal analysis cups causing excessive hardness or chill carbides

- High sulfur to manganese ratio. Reduce sulfur and/or increase manganese to stoichemetric ratio of %Mn = 1.7X %S plus 0.1% to 0.3%

Fig. 10-2. Non-uniform section, which required a dense hub, promoted chilled edges on the outer sections of the vanes.

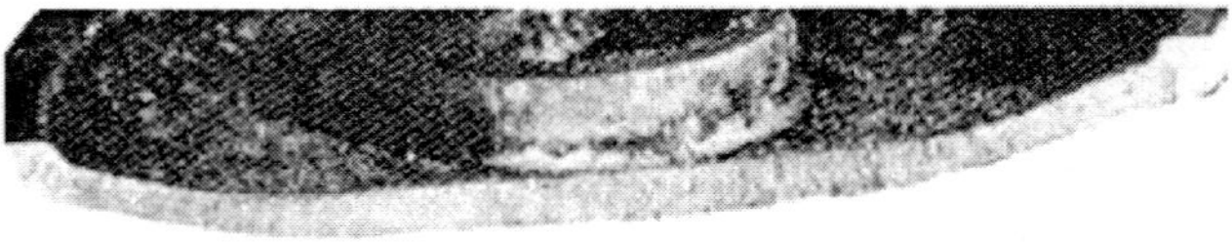

Fig. 10-3. Chilled edges caused by low carbon equivalent.

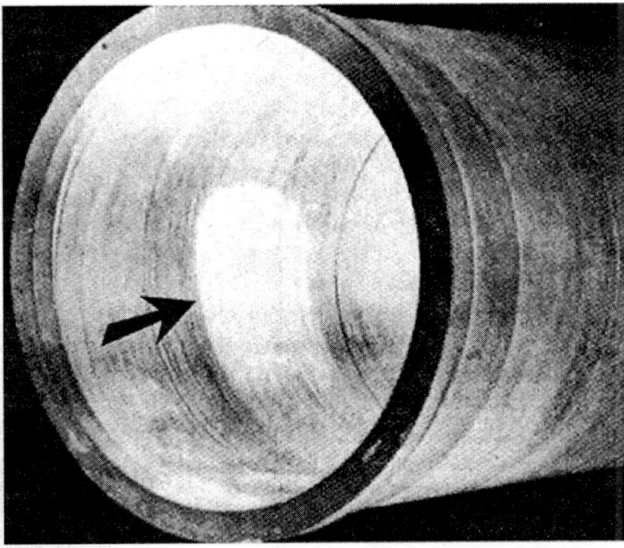

Fig. 10-4. This is a typical hard spot that became apparent during machining.

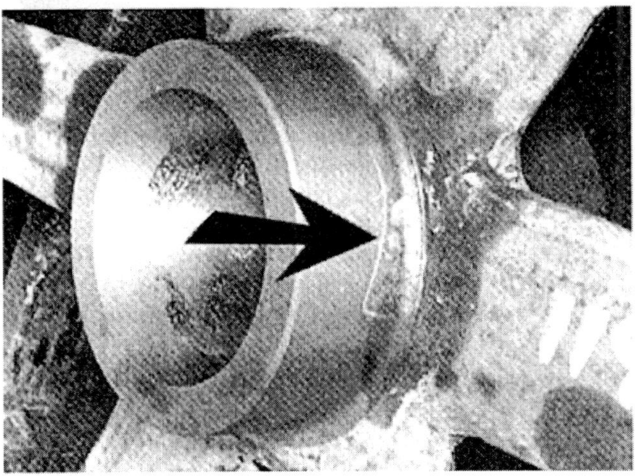

Fig. 10-5. Hard spots from undissolved ferroalloy additions are shown in the outlined area.

9. Melting Practice

- Insufficient melt temperature resulted in poor alloy distribution in the furnace, creating casting-to-casting hardness variations or variations within each casting
- Excessive superheat and oxidation cause inefficient or rapidly fading inoculation effects resulting in chill carbides
- Crucibles or ladles containing alloying element residue from prior heats create excessive hardness or chill carbides
- Incorrect inoculation causes chill carbides

  Wrong inoculant material or incompatible inoculant materials (i.e. strontium-bearing and calcium-bearing materials)

  Inoculant was added at the wrong time or excessive length of time between inoculation and pouring (fade)

  Failure to add inoculant as required including missed in-mold inoculation/treatment or misdirected in-stream inoculation devices

- Undissolved inoculant causing inclusions

  Improperly sized or improper composition of inoculant caused poor dissolution of material

  Undissolved inoculant resulting from melt temperature too low, improper dissolution time or inferior quality inoculant.

  Improper design of reaction chamber for in-mold inoculation/treatment practices, preventing incomplete dissolution of material or insufficient amount of inoculant for entire pour

  Excessive in-stream inoculation addition

10. Pouring Practice

- Cold metal as a result of cold or wet ladles
- Excessive delay in pouring
- Skull (layer of metal or dross) remained in ladles
- Wet ladles eliminated the inoculation effect

11. Miscellaneous

- Shakeout too hot (insufficient in-mold cooling time) resulting in excessive or non-uniform hardness
- Localized overheating from a cut-off torch or during grinding resulting in partial melting or matrix transformation
- Failure to heat treat before/or after welding causing wrong matrix or carbide formation.
- Wrong composition of welding rod causing wrong matrix or carbide formation
- Uneven temperature of the metal mold in the permanent mold process
- Wrong or lack of exothermic compound or sleeves for the metal poured
- Broken slag/dross filters or ceramic strainer or breaker cores causing hard inclusions

  Filters damaged during setting in mold

  Filter/gating design unable to withstand force of metal poured
- Non-uniform heating during heat treatment
- Localized overheating during heat treatment from burner impingement

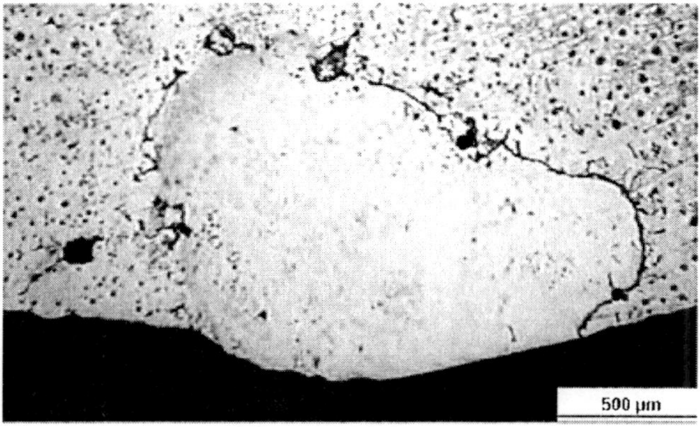

Fig. 10-6. This figure illustrates the cross-section of an undissolved inoculant inclusion in a ductile iron casting.

CHAPTER 11

Hot Tears

Description

A hot tear is a crack in the casting which occurs while it is still hot, either solid or semisolid.

Fig. 11-1. A view of a hot tear in a chilled casting caused by low phosphorus content.

Causes

1. Casting and Pattern Design

- Lack of adequate fillets
- Abrupt change in section thickness
- Any design which causes the mold to resist metal contraction

2. Pattern Equipment

- Incorrectly placed or sized cracking strips and tie bars

3. Flask Equipment and Handling Devices

- Flask bars located too near the riser or sprue
- Flask bars prevented casting contraction
- Pattern too large for the flask

4. Gating and Risering

- Gates and risers that prevent normal contraction

5. Molding Sand

- Poor collapsibility
- Low sintering-point materials that form a ceramic bond
- Sand that produces a rough surface which prevents normal contraction

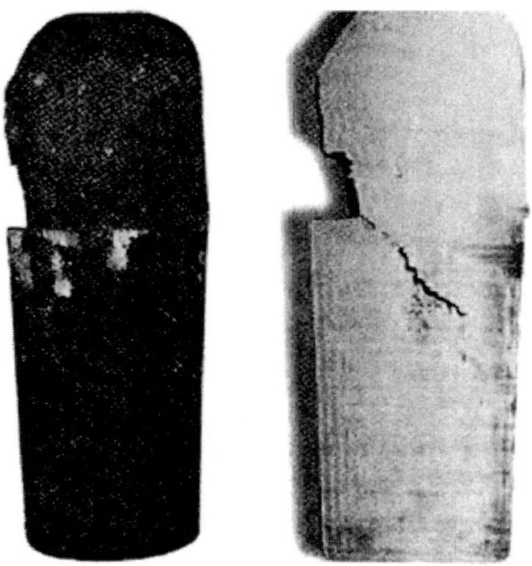

Fig. 11-2. A hot tear caused by a sharp corner.

6. Core Practice

- Poor collapsibility
- Low hot deformation with poor collapsibility (high hot strength)
- Reinforcement rods located too close to core surface
- Failure to shake out reinforced core soon enough

7. Molding Practice

- Hard ramming
- Rods, gaggers, etc., located too close to mold surface
- Riser or sprue located too close to bars or flask wall
- Excessive swabbing or sponging
- Failure to use chill when specified
- Uneven ramming caused swell and prevented contraction

Fig. 11-3. This cracked casting was caused by a core failing to collapse.

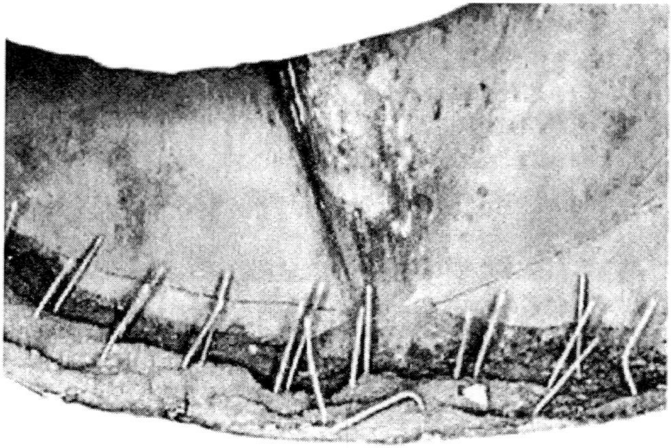

Fig. 11-4. This crack in the drag section is due to an excess amount of pitch binder in the core.

Fig. 11-5. Core rods were located too close to the casting surface.

8. Metal Composition

- Wrong composition for the design

9. Melting Practice

- Excessive carbide stabilizers
- Incorrect composition due to careless charging
- Presence of tramp elements
- Charging contaminated scrap

10. Pouring Practice

- Pouring temperature too low to promote collapsibility

11. Miscellaneous

- Shakeout too soon
- Cooling too fast after shakeout-fans at shakeout
- Spinning speed too high in centrifugal casting
- Rough flask or mold in centrifugal casting
- Rough mold transfer too soon after pouring

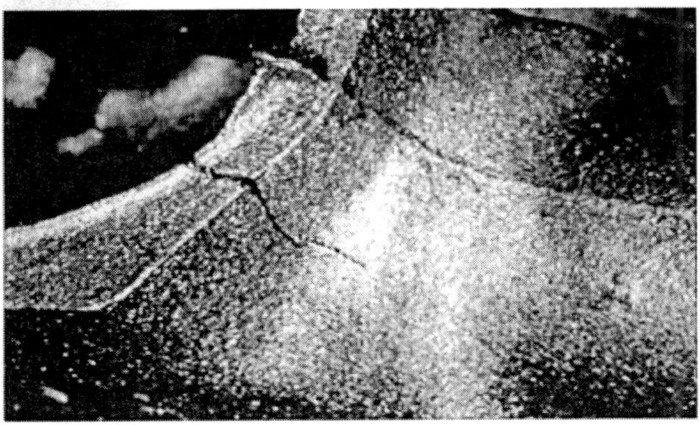

Fig. 11-6. This close-up view shows a hot crack casting defect.

Fig. 11-7. This hot tear exhibits the characteristic irregularly-shaped fracture that is located in the gate area. This defect could have been caused by a combination of a hot spot in the casting and sand discrepancies.

Fig. 11-8. This hot tear is a result of sand with too high of a dry strength.

Fig. 11-9. This is a closer view of the hot tear in Fig. 11-8.

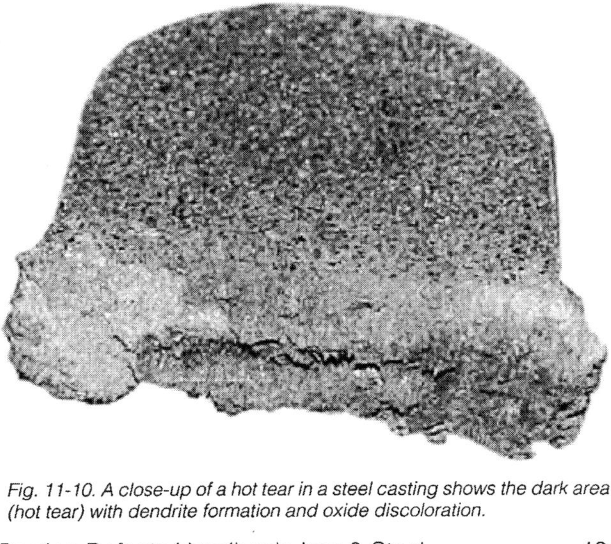

Fig. 11-10. A close-up of a hot tear in a steel casting shows the dark area (hot tear) with dendrite formation and oxide discoloration.

CHAPTER 12

Inverse Chills

Description

Inverse chill defects are also known as reverse chills. This condition occurs in gray iron, ductile iron and malleable iron castings. Casting sections where the interior is chilled or white, while the outer sections are mottled, gray or ductile iron.

Fig. 12-1. This inverse chill on a test bar was caused by cold metal and wet sand.

Causes

1. Casting and Pattern Design

- Thin sections adjacent to heavy sections in malleable iron

2. Pattern Equipment

- None

3. Flask Equipment and Handling Devices

- None

4. Gating and Risering

- None

5. Molding Sand

- Excessively wet sand causing hydrogen contamination of the molten metal

6. Core Practice

- Any practice involving excessive moisture

  Improperly dried core wash or adhesive

  Absorption of moisture in susceptible sand binder systems.

  Incomplete purging of amine catalyst gases.

Fig. 12-2. Inverse chill caused by cold metal and wet sand.

7. Molding Practice

- Any practice involving excessive moisture.
  - Improperly dried mold wash or adhesive
  - Absorption of moisture in susceptible sand binder systems.
  - Incomplete purging of amine catalyst gasses.

8. Metal Composition

- The presence of nonferrous metals such as aluminum or titanium that increases hydrogen absorption in molten cast iron
- Carbide stabilizers, such as tellurium, bismuth, chromium, etc.
- Sulfur content not balanced with manganese in ratio of %Mn =1.7X%S plus 0.1% to 0.3%
- High titanium levels in combination with low sulfur level.
- In ductile iron, magnesium in excess of that required to provide good nodularity

9. Melting Practice

- Any condition which would allow or lend itself to the dissolution of free hydrogen gas in the metal
  - Hydrogen atmosphere from low coke bed in the cupola
  - Wet linings in cupola, electric or arc furnace
  - Use of contaminated metals in charge, such as die cast non-ferrous metals including aluminum or titanium.
  - Boiling metal caused by rusty charge material
  - Wet bottom sand in the cupola
- Inadequate inoculation or excessive holding time between inoculation and pouring (fade).
- Excessively superheated or excessively long holding times prior to inoculation and pouring.

10. Pouring Practice

- Wet ladles or ladle lips

11. Miscellaneous

- None

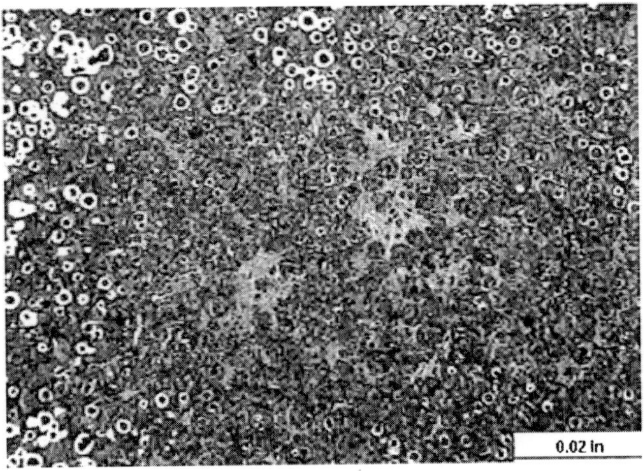

Fig. 12-3. An inverse chill in pearlitic ductile iron

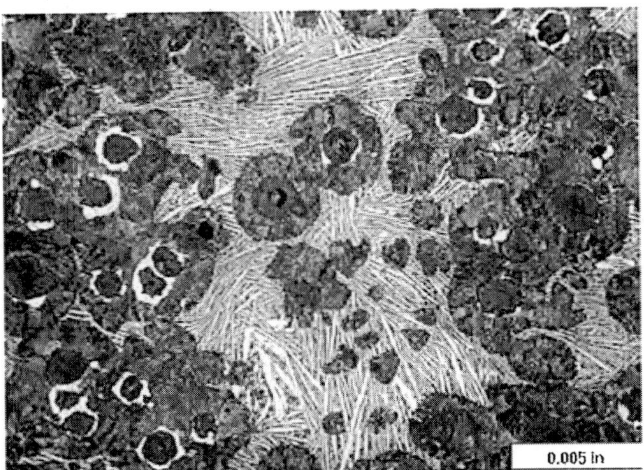

Fig. 12-4. An inverse chill in a pearlitic-ferritic ductile cast iron is shown.

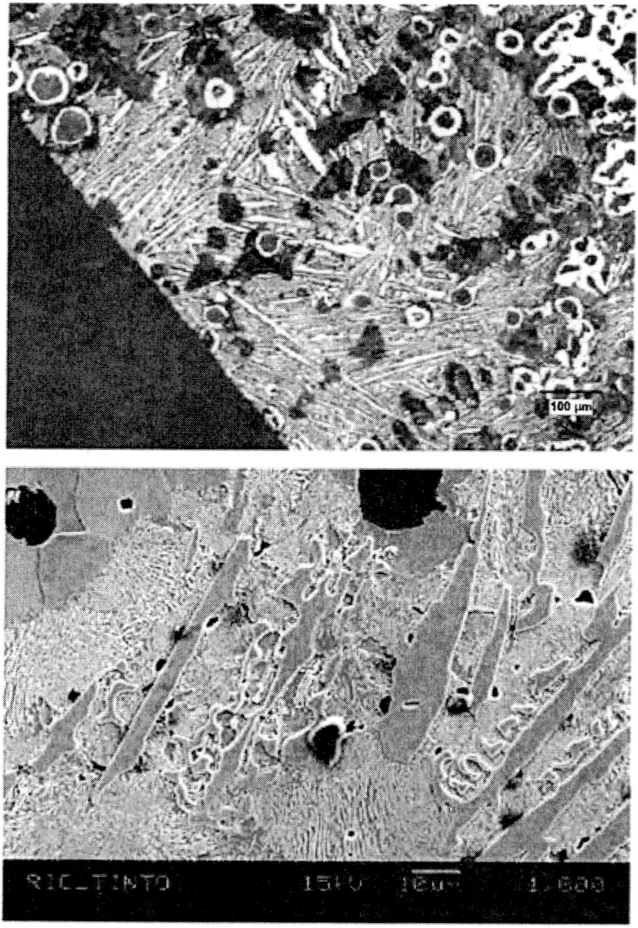

Fig. 12-5. Two views of an inverse chill; the metallographic view on left and SEM view on the right.

CHAPTER 13

Mass Hardness

Description

Mass hardness is a condition in which the hardness values for the entire casting are too high for the application.

Causes

1. Casting and Pattern Design

- None

2. Pattern Equipment

- None

3. Flask Equipment and Handling Devices

- None

4. Gating and Risering

- None

5. Molding Sand

- Wet sand in combination with off analysis
- Use of other material, such as zircon sand or chromite sand that has a more rapid transfer of heat

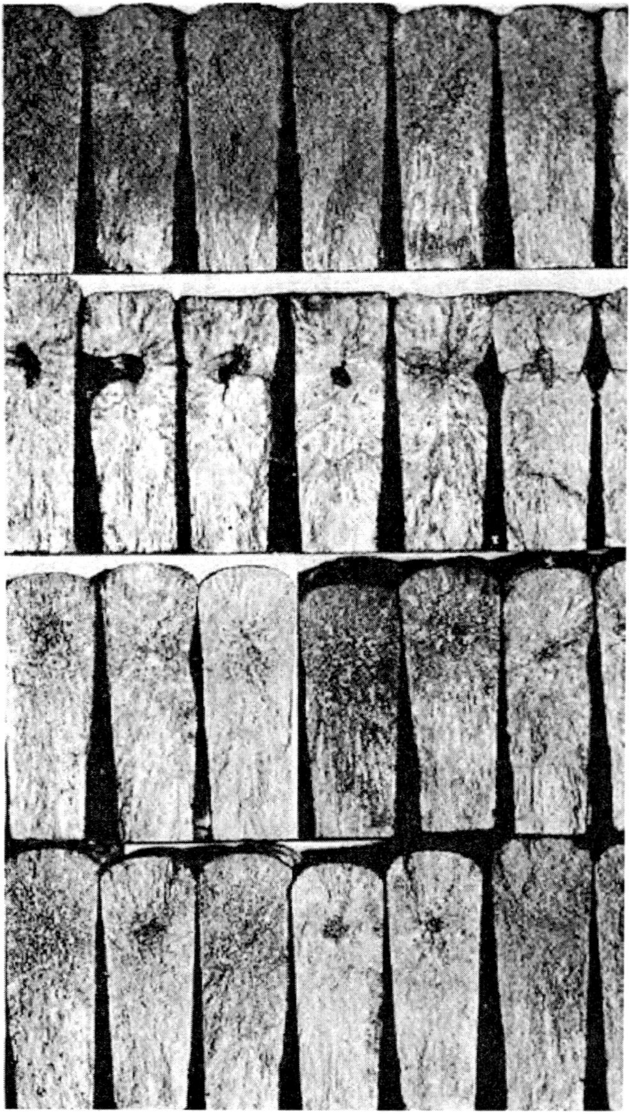

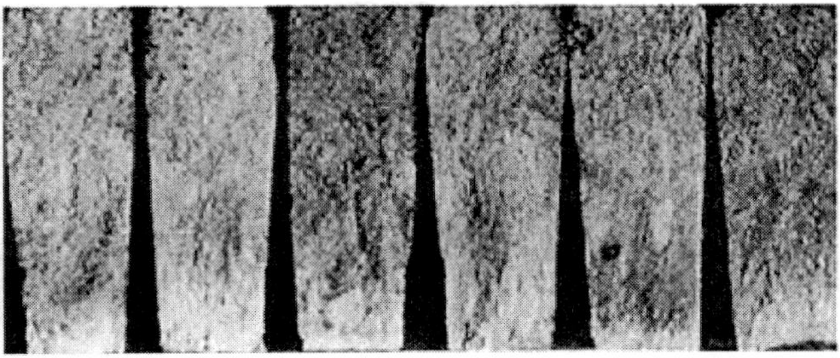

Fig. 13-1. Chill sequence indicates occurrences of charge material contamination when stainless steel scrap was inadvertently charged instead of regular steel scrap.

6. Core Practice

- Excessive use of chill coat
- Too thin a layer of sand over arbors

7. Molding Practice

- Improper use of chill techniques
- Misapplied chill washes
- Low temperature of the molds in die casting or permanent molds
- Inadequate insulation in permanent mold

8. Metal Composition

- Incorrect metallurgy
- Steel-excessive carbide stabilizing alloys
- Gray and ductile iron
  - Too low carbon equivalent
  - Tramp element
  - Excessive carbide stabilizing elements
- Inadequate or inappropriate inoculation in thin sections
- Malleable-tramp elements that prevent nodulizing or malleablizing

9. Melting Practice

- Off analysis
- Oxidizing melting conditions
- Super heating of most metals
- Careless charging
- Low coke bed in cupola

10. Pouring Practice

- Oxidized metal
- Wet, dirty and contaminated ladle spouts and runners
- Pouring too cold in thin sections

11. Miscellaneous

- Air quench due to early shakeout
- Improper heat treatment
- Over chilling with water-cooled permanent molds
- Unintentional water quench

CHAPTER 14

Metal Penetration and Fusion

Description

Metal penetration is a condition in which the metal or metallic oxides have filled the voids between the sand grains without displacing them or by chemically changing the silica or binder. In conventional molds or cores this can take place due to an open grain or surface porosity. In high density molds, iron oxide readily combines with the silica to form a fayalite which is fluid at several hundred degrees lower temperature than that of the original metal. Generally, the higher the free moisture content at the metal sand interface the greater will be this phenomena.

Fig. 14-1. This penetration defect in a steel casting was due to poor compaction on the vertical walls.

Fusion is a related surface defect having a rough glossy appearance resulting from metallic oxides acting as a flux on the sand or low fusion binders which melt leaving voids on the surface. Controlled fusion may result in a hard cake or fused layer which peels away from the casting leaving a smooth finish.

Causes

1. Casting and Pattern Design

- Sharp corners
- Overhanging or protruding sections
- Thin core or sand section surrounded by heavy metal

2. Pattern Equipment

- Pattern equipment constructed so as to promote uneven or soft ramming
- Improper location of parting line which induces soft ramming
- Improper location of the pattern on the plate causing narrow pockets
- Multiple patterns mounted too close together
- Improper location of sprues, risers and runners
- Inadequate draft
- Pattern equipment constructed to promote excessive metal pressure such as a vertical rather than horizontal mounting.
- Inadequate provision for mold and core venting in high density molding
- Different squeeze pressure between the cope and drag.

Fig. 14-2. This defect is a result of different squeeze pressure between the cope and drag.

3. Flask Equipment and Handling Devices

- Design and construction of flask equipment which promotes uneven or soft ramming
- Excessive cope height
- Lack of venting of flask and bottom boards in high density molding
- Metal inserts or chills which act as condensers for moisture

4. Gating and Risering

- The location of gates and risers which promote localized overheating of the sand
- Incorrect ratio of sprue, runner, and gate permitting metal oxidation during pouring
- Pouring cup or sprue designed so as to permit aspiration of air
- A riser neck which overheats the sand

Fig. 14-3. Photomicrograph showing the type of metal structure often found between sand grains and at a considerable distance from mold/ metal interface; (original photo 400X).

5. Molding Sand

- A low sintering or low fusion sand or binder
- Any sand condition that promotes nonuniform density
- Improper distribution of the sand grains which induce poor ramming
- Poor flowability or moldability caused by: moisture segregation; poorly mixed sand; excessive moisture; excessive cereal; excessive green deformation; and prepacked or lumpy sand
- Nonuniform coating of shell sand
- Insufficient carbonaceous or deoxidizing material which will prevent the water vapor from reacting with the metal
- Excessive mold permeability or, in the case of high density molding, insufficient permeability

6. Core Practice

- Too low a sintering or fusion point of sand or binders
- Grain size which is too large or improperly distributed in uncoated cores
- Soft rammed cores
- Plugged vents
- Improperly tucked sand under loose pieces
- A rough or porous core surface
- Core surface which is bruised or abraded
- Improper or insufficient release agent
- A partially disintegrated core surface due to overbaking
- A broken or disturbed core wash due to rough handling
- A dirty core box
- In self curing binders; improper ratio of catalyst to binder

- Poorly mixed core sand
- Any condition promoting veining or finning such as a core crack
- High cereal, low core oil, or low moisture
- Low hot compressive strength
- Poorly cleaned or finished cores
- Dipped or sprayed cores that are not redried
- Cores moisture increased during storage
- Low air pressure for core blower or rammer which results in soft-rammed cores

Fig. 14-4. A typical hard mold penetration caused by using a green sand mold rammed to 85-90 hardness.

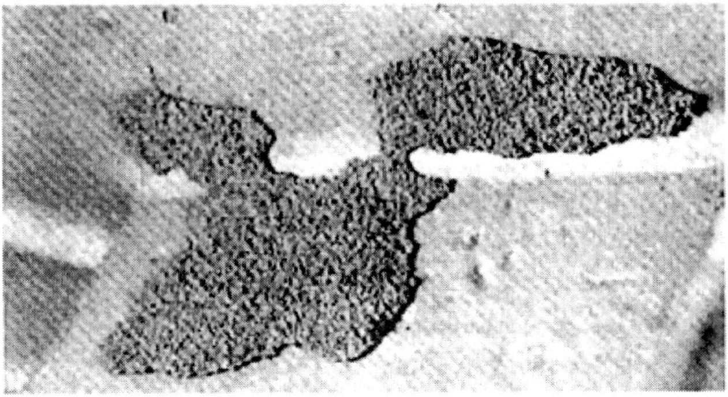

Fig. 14-5. Insufficient drying of mold wash created a penetrated area.

7. Molding Practice

- Soft or uneven ramming
- Excessive ramming or density
- Poorly patched areas
- Loose fitting stop-off
- Uneven or insufficiently dried mold coatings
- Excessive use of parting compounds
- Insufficient facing to cover the pattern
- Low air pressure on molding machine or rammer
- Any combination of hot and cold materials which can cause condensation
- Combining high permeability facing sand with a low permeability backing sand particularly in high density molding

8. Metal Composition

- Metals which form fluid oxides
- Low melting constituents such as lead or tin
- Metal oxides which readily react chemically with sand
- Metals requiring high pouring temperature

9. Melting Practice

- Oxidized metal
- The excessive use of any fluxing-type material such as limestone, soda ash, and fluorspar
- Melting too hot

10. Pouring Practice

- Excessively high pouring temperature
- Excessive pouring height
- Poor skimming practice

11. Miscellaneous

- Any factor involving free moisture in the mold cavity
- Shakeout too cold or too late, particularly in high density molding
- Sand coated with low fusion materials
- Sand or binder reacting chemically with metal oxides
- Insufficient venting in high density molding

Fig. 14-6. Gray iron engine block casting with (bottom) and without (top) metal penetration resulting from a soft green sand ram.

CHAPTER 15

Misruns and Cold Shuts

Description

A misrun casting lacks completeness due to failure of the metal to fill the mold cavity. There may be a smooth rounded-edge hole through the casting wall. One or more extremities may be only partially filled.

A cold shut casting has a definite discontinuity due to imperfect fusion where two streams of metal have converged. This defect may have the appearance of a crack or seam with smooth, rounded edges.

Both defects may be related to insufficient fluidity of the metal for the section of casting being teemed or excessive gas pressure resisting the flow of liquid metal into the mold. It is advisable to check the chapter on *Gas Defects* before assuming that cold metal is the only cause.

Causes

1. Casting and Pattern Design

- Non-uniform casting sections resulting in interrupted metal flow
- Metal sections too thin and of too great expanse for the fluidity of the metal poured
- Isolated thin sections

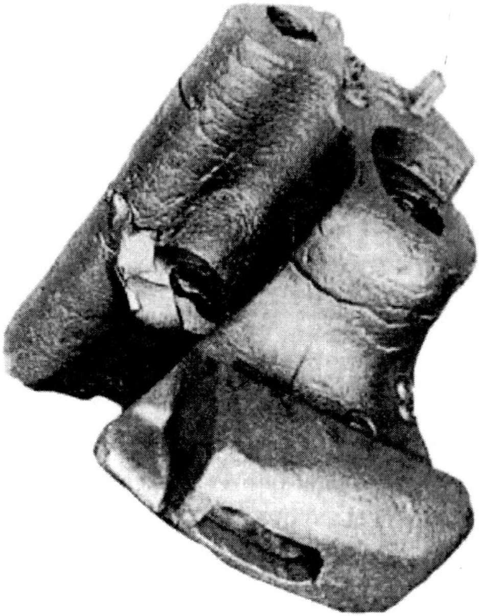

Fig. 15-1. Laps or cold shuts caused by the pourer missing the pouring cup for a few seconds.

2. Pattern Equipment

- Worn patterns or coreboxes resulting in cavities with thin metal sections
- Patterns not made according to blueprint
- Pattern equipment not properly reinforced
- Misalignment of pattern or core equipment
- Permeability of the lost foam pattern
- Inconsistency in foam pattern quality
- Foam pattern breakage
- High foam pattern density
- Unacceptable foam pattern refractory coating
- Excessive foam pattern glue

Fig. 15-2. A typical misrun, here one tooth is completely missing.

3. Flask Equipment and Handling Devices

- Cope or drag shift resulting in metal sections that are too thin
- Worn flask pins or bushings
- Patterns not properly mounted
- Failure to tilt molds caused interrupted metal flow due to insufficient pressure
- Flask insufficiently barred allowing cope to sag
- Distorted or dirty jackets caused crush
- Weak or improperly reinforced flasks
- Distorted bands
- Cope flask too shallow resulting in insufficient liquid metal head pressure

4. Gating and Risering

- Improper sizing of sprue, runners, or ingates. Ingates too small or of insufficient number
- Gates not properly located to feed thin sections
- Improper distribution or balance of gates
- Liquid metal head pressure was too low

5. Molding Sand

- Excessive moisture
- Sand composition too high in volatile matter
- Sand too weak allowing mold distortion
- Permeability too low resulting in excessive pressure in the mold cavity
- Heat transfer of molding material too rapid, reducing the metal temperature (fluidity) too quickly
- Excessive carbonaceous additive (lustrous carbon) and binder

Fig. 15-3. Typical cold shut on an impeller casting, the defective vane is broken.

Fig. 15-4a & b. Two views of lustrous carbon defects, a distinctive form of cold shut with a line of demarcation in the metal having a carbonaceous film.

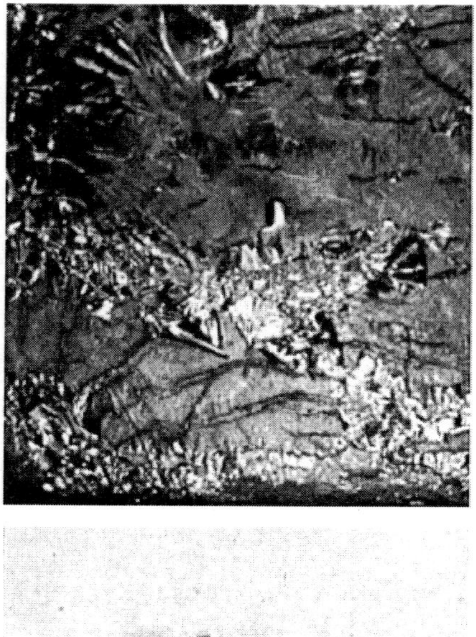

Fig. 15-4c & d. Closer views of lustrous carbon.

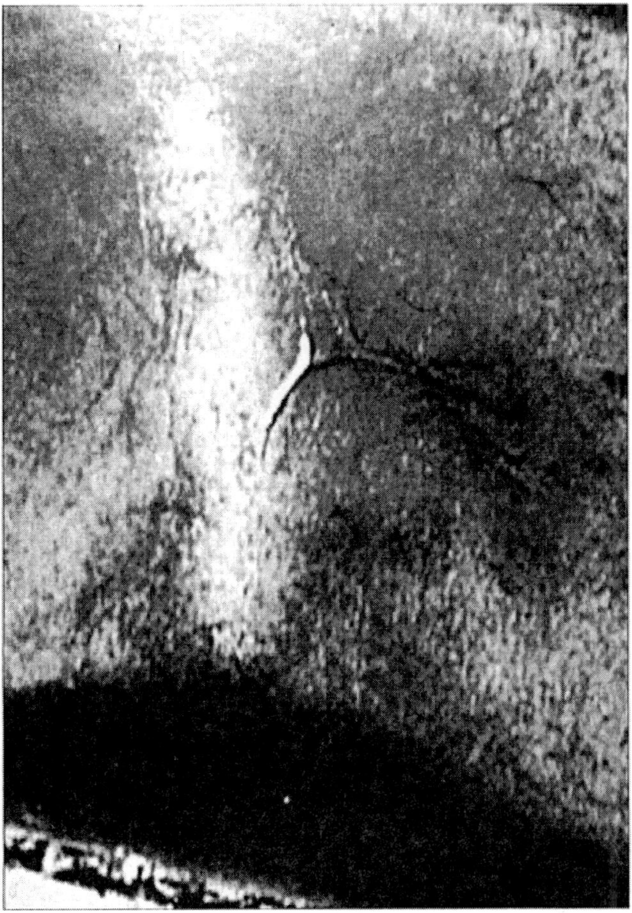

Fig. 15-5. A casting exhibiting a cold shut defect.

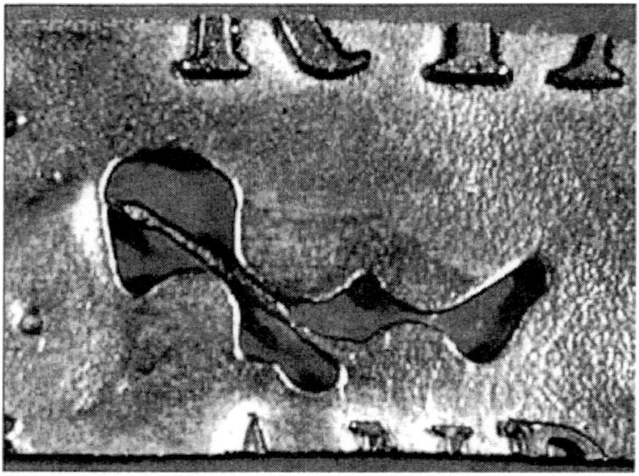

Fig. 15-6. This casting shows a misrun, a lack of completeness due to a failure of the metal to fill the mold cavity.

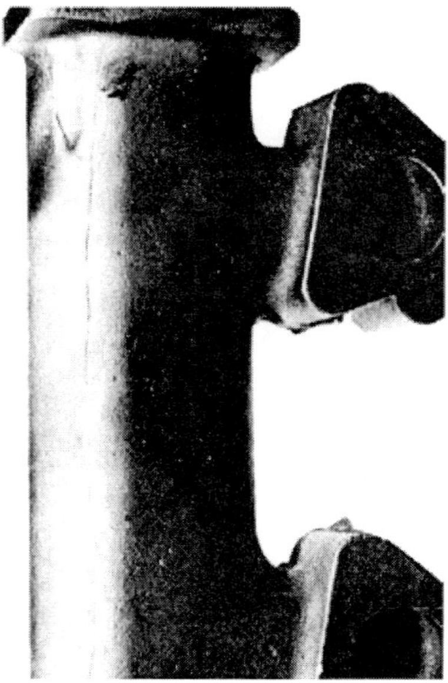

Fig. 15-7. A cold shut caused by small ingates.

6. Core Practice

- Excessive organic core binders
- Undercured or green cores
- Insufficient permeability
- Insufficient vents resulting in excessive back pressure in the mold cavity
- Cores not properly sized or fitted allowing decreased metal sections
- Excess wash, mudding compound or paste. Metal sections may be reduced and/or excess gas evolved
- Raised or sagged cores
- Core shift causing too thin metal sections
- Cores warped or distorted-thermoplastic binders
- Reinforcing rods too close to core surface resulting in chilling effect
- Cores with too rapid heat transfer characteristics chill the metal.
- Add 0.5% to 1% iron oxide (lustrous carbon)

7. Molding Practice

- Excessive ramming causing low mold permeability
- Soft ramming permitting mold distortion or sagging
- Mold insufficiently reinforced
- Improperly bedded or clamped bottom boards
- Excessive sticking or patching
- Molds for thin flat castings set level instead of tilting
- Too few or too small chaplets allowing core to rise
- Coreprints too tight
- Excessive wash
- Misalignment of sprue
- Insufficient venting

8. Metal Composition

- Gray, malleable or nodular irons with too low of carbon equivalent
- Low phosphorus in gray iron
- In steel composition changes affect fluidity

Fig. 15-8. Cold metal poured during the first part of the heat caused poor definition of both edges and cored holes.

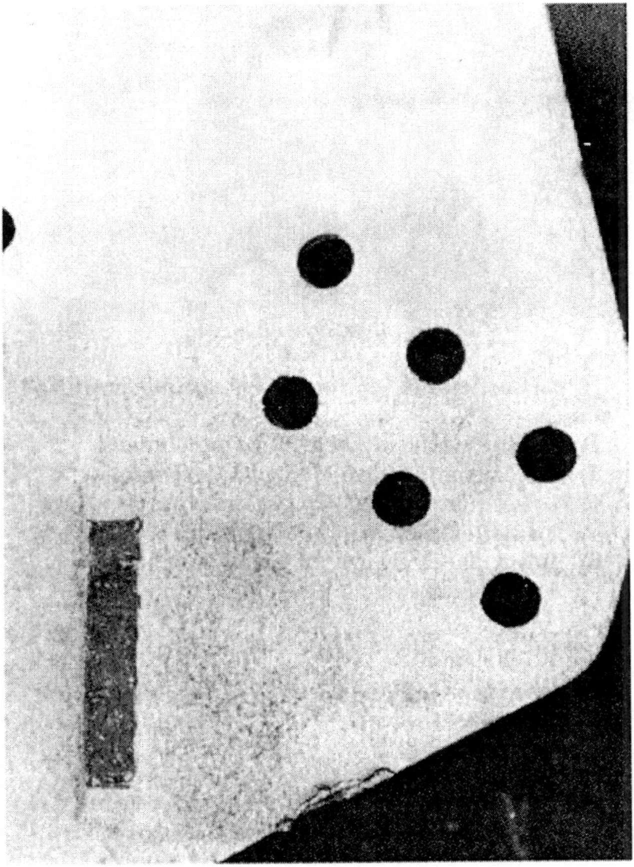

Fig. 15-9. Same casting as in previous figure after correction.

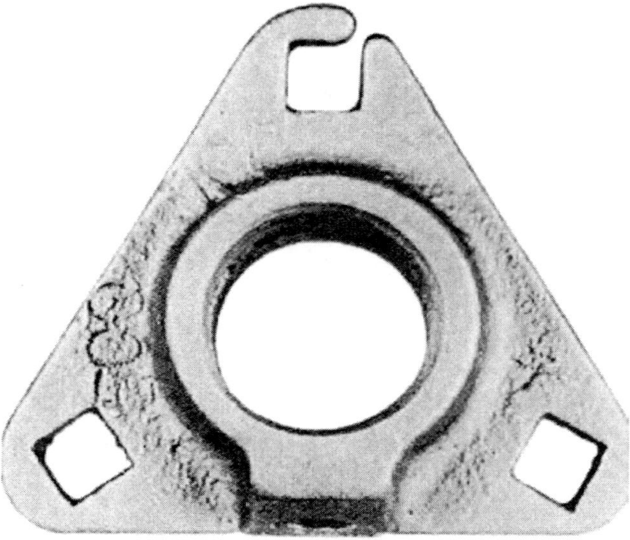

Fig. 15-10. A misrun in this casting was caused by cold metal.

9. Melting Practice

- Wrong composition due to careless weighing or charging
- Cold melting
- Oxidized or gassy metal
- Over reduced metal, hydrogen reduces fluidity
- Excessive ladle additions
- Damp ladle additions

10. Pouring Practice

- Pouring temperature too low
- Interrupted pouring
- Reducing the rate of pour too soon
- Boiling metal from wet spout, wet lining, etc.
- Cold ladles
- Thin castings poured on the level
- Low head pressure
- Slag, dross or ladle refractory which plugs the gate
- Poured short
- Pouring too slowly

11. Miscellaneous

- Chills and chaplets too large for metal section
- Reduced metal section from any cause, i.e., excessive mold weights, careless weight shifting, or misalignment of mold halves
- Condensation from warm sand on cold cores, chills, inserts, or chaplets

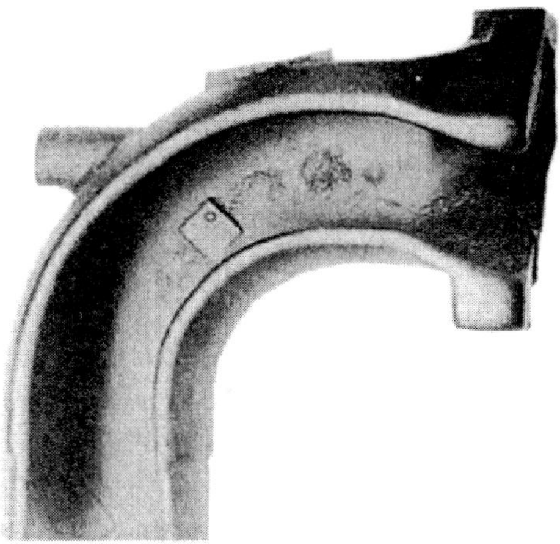

Fig. 15-11. A misrun caused by gates so small that the gates froze before the casting was filled.

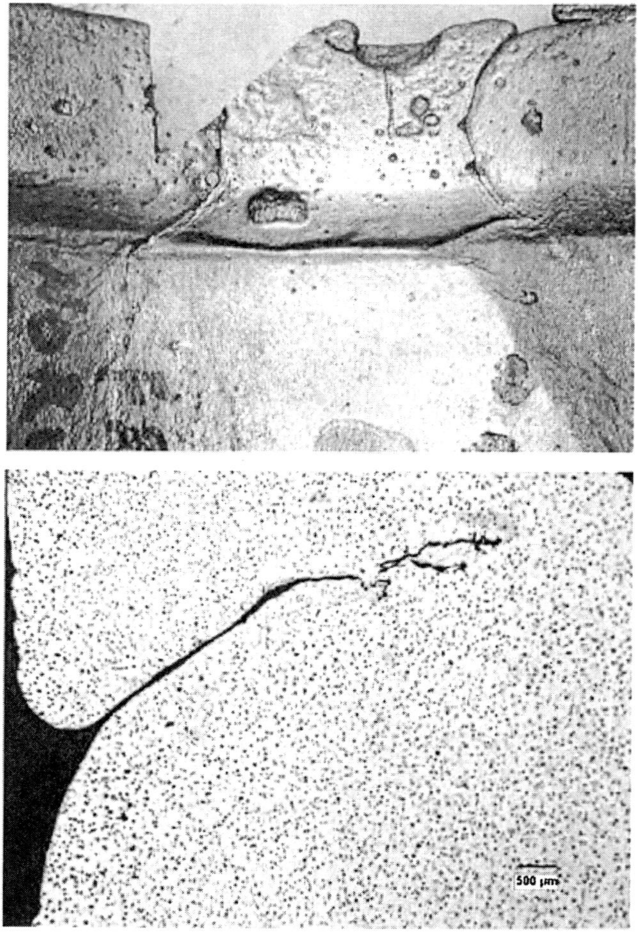

Fig. 15-12. These two casting photos show a misrun (top) and a dross defect (bottom).

CHAPTER 16

Off Dimension Related To Cores

Description

An off dimension defect is an incorrect dimension occurring as a result of the wrong core being used, correct cores improperly set, cores omitted, cores incorrectly assembled, or thermal instability of cores or mechanical damage to the pattern.

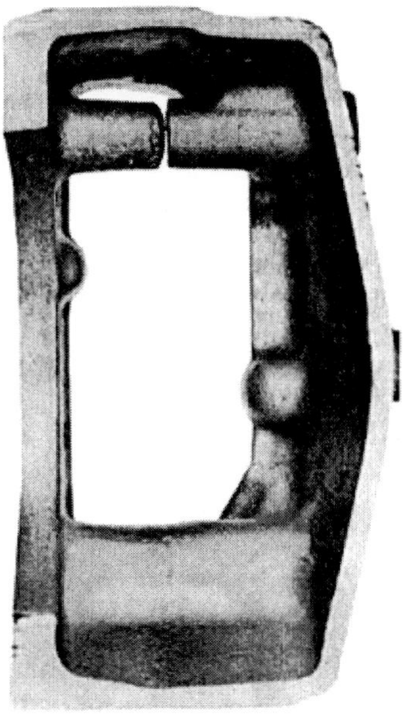

Fig. 16-1. This off dimension defect was a result of failure to clean the core assembly prior to setting the mold (boss was intended to be solid).

Causes

1. Casting and Pattern Design

- Incorrect pattern design allows core flotation
- Internal horizontal channels with large length to thickness ratio in casting causing core bending or fracture
- Similarity of coreprints between similar parts allows "wrong core" error

2. Pattern Equipment

- Unidentified locators or prints
- Improper or broken telltales
- Improperly marked coreboxes and patterns
- Faulty corebox construction, causing distortion
- Coreprints too large or too small
- Chaplet location not marked
- Chaplet size not marked
- Did not use core assembly/setting fixture
- Mechanical damage to the pattern

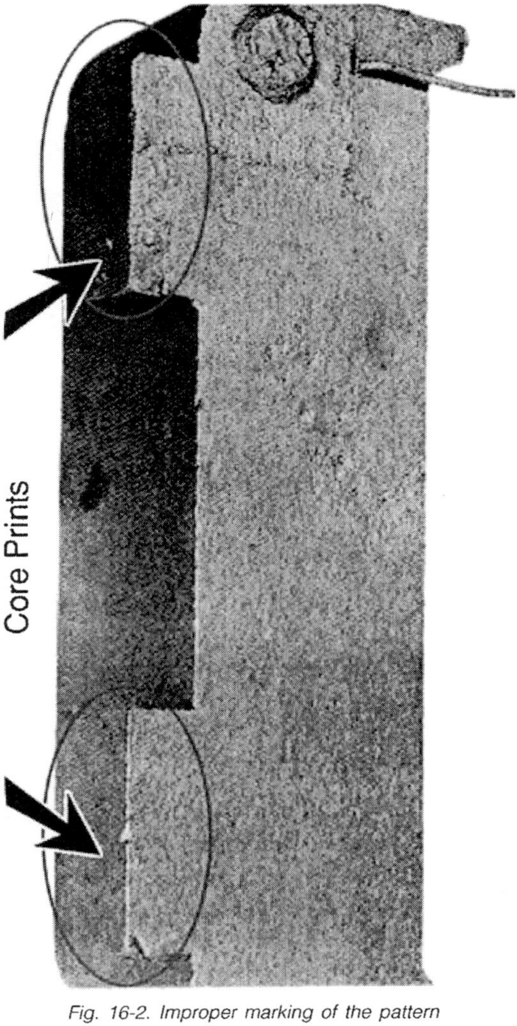

Fig. 16-2. Improper marking of the pattern caused the molder to omit placing the required cores.

Fig. 16-3. Core was omitted because print was incorrectly identified.

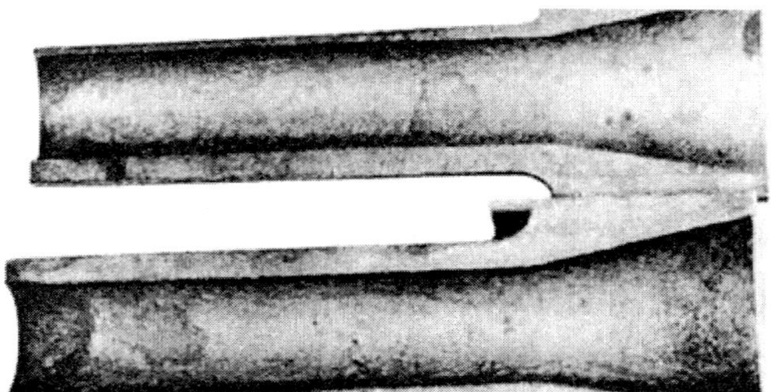

Fig. 16-4. Excessive metal pressure caused the core to warp during pouring.

Fig. 16-5. An oversized coreprint permitted metal flow around the entire core.

3. Flask Equipment and Handling Devices

- Coresetting jig or fixture worn or inaccurate

4. Gating and Risering

- None

5. Molding Sand

- None

6. Core Practice

- A wrong core was used
- Cores improperly set
- Sagged or distorted cores
- Too much or lack of paste
- Wash coating too thick or too thin
- Improper rodding
- Conversion of oil sand core to self-cure core resulting in lack of compensation for stability
- Undercured cores
- Improper grain distribution
- Too rapid heat transfer of cores
- Improper core storage time and conditions (temperature, dust, humidity, transportation without protection)
- Carelessness in coremaking and/or core assembly

7. Molding Practice

- Setting defective cores
- Improper filing or rubbing of cores

8. Metal Composition

- Mottled structure in malleable iron will distort and grow when heat treated

9. Melting Practice

- None

10. Pouring Practice

- Pouring too hot
- Pouring too hard

11. Miscellaneous

- None

CHAPTER 17

Open Grain Structure

Description

Open grain structure is a condition wherein a casting, when machined or fractured is too coarsely grained. This structure may occur throughout the entire casting or it may be localized. The most important areas of control are in design, gating, composition and melting. This is primarily considered a metallurgical-type defect. It is often difficult to differentiate between open grain, porosity and microshrinkage defects.

Fig. 17-1. This open grain structure was caused by an excessively high carbon equivalent.

Causes

1. Casting and Pattern Design

- Slow cooling in heavy sections
- Lack of cores in heavy sections that are detected during machining operations
- Sections that promote heat retention
- Lack of liaison between the designer, patternmaker, foundry and machine shop

2. Pattern Equipment

- Improperly cored heavy sections (See *Section 1-* above)

3. Flask Equipment and Handling Devices

- Lack of rigidity in flasks

Fig. 17-2. This iron casting exhibits an open grain structure.

4. Gating and Risering

- Gating which does not permit directional solidification
- Ingate and riser locations create hot spots in localized areas
- Oversized gates and risers
- Insufficient number of ingates

5. Molding Sand

- High deformation in sand, permitting mold wall movement
- Improper selection of molding additives
- Molding materials that retard heat transfer

6. Core Practice

- Undercured cores
- Exothermic materials or insulating materials

7. Molding Practice

- Lack of chills or chill promoting materials

8. Metal Composition

- Improper analysis for section size (i.e., carbon equivalent too high for the section size, permitting Type C graphite to form.)
- Gray iron and malleable-carbon equivalent or phosphorus levels too high
- Insufficient carbide and or pearlite stabilizers
- Excessive use of graphitizers

9. Melting Practice

- Cupola melted gray iron

  Improper balance of air and coke creating an excessive carbon pickup

  High coke bed

  Slow tapping out or excessive shutdowns

  Careless charging or weighing of materials

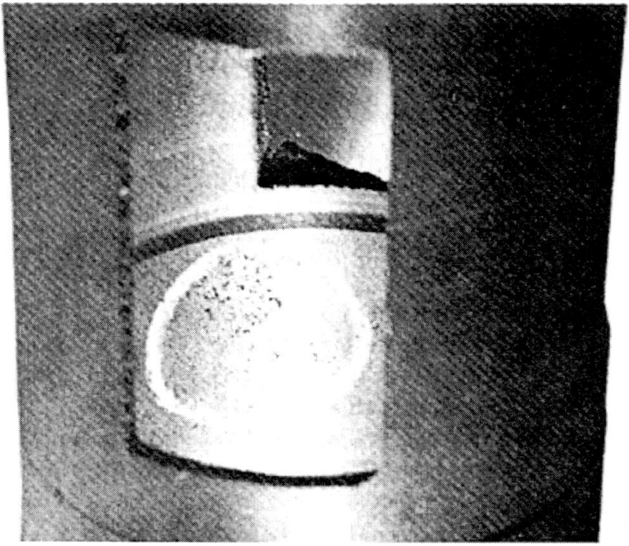

Fig. 17-3. This area (in circle) shows an open grain structure that was caused by a high phosphorus content.

10. Pouring Practice

- Pouring too hot

11. Miscellaneous

- Improper heat treatment (i.e., any heat treatment that promotes graphitization)
- Improper machining (i.e., dull tools, excessive depth of cut, excessive cutting speeds)

CHAPTER 18

Poured Short

Description

Castings lack completeness due to an under-filled mold. The upper section may be under-filled or even missing. Casting surfaces adjacent to the area may have lost definition and be rounded in appearance.

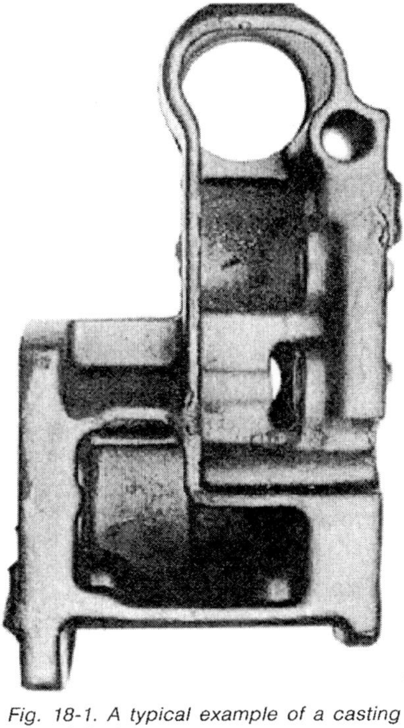

Fig. 18-1. A typical example of a casting poured short.

Causes

1. Casting and Pattern Design

- None

2. Pattern Equipment

- None

3. Flask Equipment and Handling Devices

- None

4. Gating and Risering

- Inadequate metal flow
- Filter blockage
- Improper filter sizing
- Improper filter selection

5. Molding Sand

- Permeability too low, resulting in excessive pressure in the mold cavity
- Sand too high in volatile matter

6. Core Practice

- Excessive organic core binders
- Undercured or green cores
- Insufficient permeability
- Insufficient number of vents resulting in excessive back pressure in the mold cavity.

7. Molding Practice

- Excessive ramming causing low mold permeability
- Improperly bedded or clamped bottom boards
- Insufficient number of vents resulting in excessive back pressure in the mold cavity.

8. Metal Composition

- None

9. Melting Practice

- None

10. Pouring Practice

- Insufficient metal
  - Inaccurate melt/ladle weight scales
  - Leaking ladle
- Improper pouring technique
- Cold metal freezing in gate

11. Miscellaneous

- Inadequate metal flow
- Filter blockage
- Improper filter sizing
- Improper filter selection

Fig. 18-2. An additional example of a casting poured short.

CHAPTER 19

Ramoffs or Ramaways

Description

A ramoff or ramaway is an incorrect casting dimension resulting from the sand moving away from the pattern during molding.

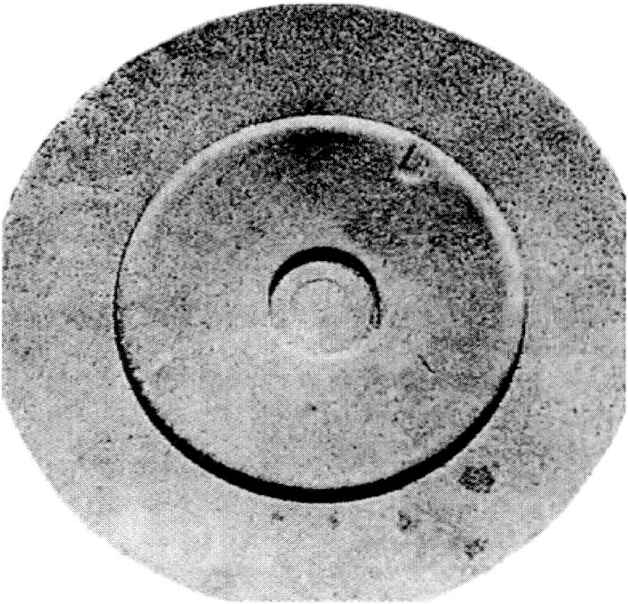

Fig. 19-1. This casting hub section shows a typical ramoff.

Causes

1. Casting and Pattern Design

- A curved or rounded vertical surface
- Extended vertical sections
- Sharp edges
- Lack of proper fillets or radii

2. Pattern Equipment

- Loose pieces not properly secured to pattern
- Patterns not equipped with ramoff strips
- Patterns located too close together
- Patterns located too close to edge of flask
- Patterns and plates too smooth

3. Flask Equipment and Handling Devices

- Loose squeeze head on machine
- Loose pattern plate on a jolt machine
- Inadequate support under the pattern
- Incorrect design or placement of squeeze blocks
- Weak bottom boards
- Molding machine worn or out of level

Fig. 19-2. Result of peening in only one direction, causing the round boss to become elliptical.

4. Gating and Risering

- A gate or riser located too close to the pattern

5. Molding Sand

- Sand characteristics that contribute to non-uniform ramming will aggravate any of the design or mechanical conditions

  Excess moisture

  Improper mulling

  Excess cereal

  Low green compressive strength, for the application

6. Core Practice

- Ramming or peening in the wrong direction
- Improper tucking around loose pieces and bosses
- Lack of proper venting of core boxes (in blown cores)
- Improper spacing or location of blow tubes
- Core sand mixture not applicable to the job

7. Molding Practice

- Ramming or peening in the wrong direction
- Improper tucking around loose pieces and bosses
- Whipping of a pattern due to worn machine
- Pre-ramming of sand while filling flask
- Improper filling of flask with sand
- Overfilling flask before peening
- Not leveling the sand in flask before ramming or jolting

8. Metal Composition

- None

9. Melting Practice

- None

10. Pouring Practice

- None

11. Miscellaneous

- Rollover machines worn or out of adjustment
- Pattern draw equipment worn or out of adjustment
- Improper flow or direction of sand into flask

CHAPTER 20

Rough Surface

Description

The casting surface lacks the required degree of smoothness for a specific application.

Fig. 20-1. This typical rough surface was caused by an uncoated open core.

Causes

1. Casting and Pattern Design

- Design is a contributing factor only when sharp corners, deep pockets and the like are involved

2. Pattern Equipment

- Inaccessible areas
- Fillets too small
- Rough pattern surface
- Weak patterns
- Materials and coatings that promote sticking of sand
- Nonuniform heating in shell or hot box

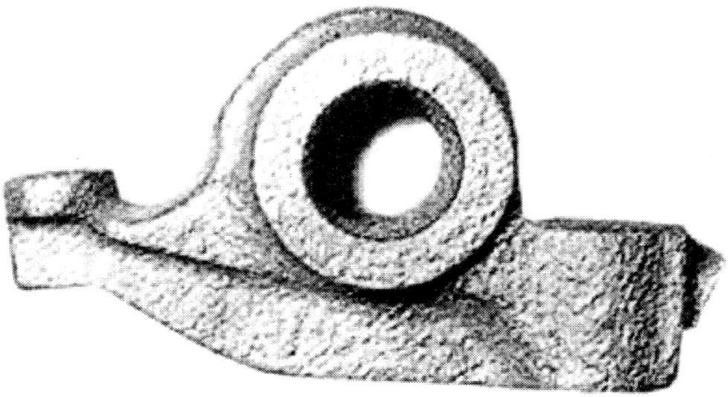

Fig. 20-2. Excessive moisture in molding sand created a low moldability and an open mold surface.

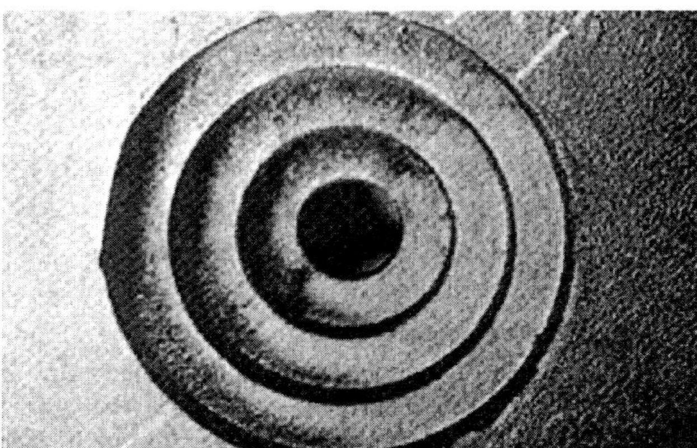

Fig. 20-3. An unwashed (upper left section) and a washed section (lower right) of a mold are shown.

3. Flask Equipment and Handling Devices

- Flask bars located too close preventing proper ramming
- Too small a flask

4. Gating and Risering

- Hot spots created by the gating
- High metal velocity at ingates
- Excessive head pressure
- Poor gating required a higher pouring temperature

5. Molding Sand

- Moisture too high
- Moisture too low
- Permeability too high-coarse sand
- Improper mulling
- Foreign material
- Low fusion point materials
- Low carbonaceous materials
- High carbonaceous materials
- Poor flowability
- Insufficient hot compressive strength
- Carbonaceous materials too coarse
- Excessive use of release agents
- Insufficient new sand additions

6. Core Practice

- Dirty core boxes
- Soft rammed cores
- Improper application of core wash
- Over or undercured cores
- Rough handling
- Permeability too high-coarse sand
- Hot strength too low
- Improper mixing
- Poorly cleaned and finished cores
- Incorrect release agent

7. Molding Practice

- Soft ramming
- Dirty patterns
- Improperly riddled or screened sand
- Improper application of mold coatings
- Poorly finished molds
- Excessive use of parting compounds

8. Metal Composition

- Composition requires a high pouring temperature
- Composition reacts to the mold surface

9. Melting Practice

- Poor slag practice

10. Pouring Practice

- Too high pouring temperature
- Improperly skimmed ladles
- Ladle too high

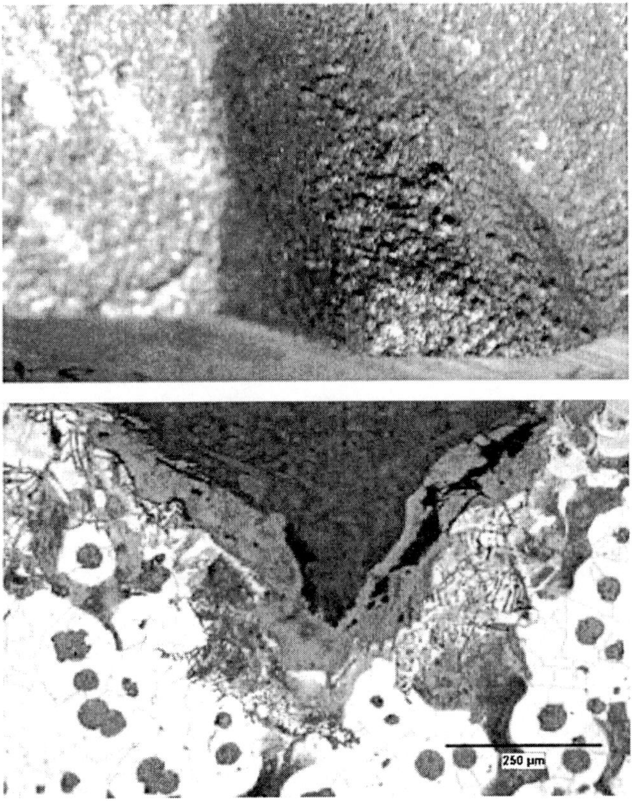

Fig. 20-4. Surface defects due to a mold metal reaction (top). A cross section of the surface shown in figure on left detected iron oxide in the cavities (bottom).

11. Miscellaneous

- Improper cleaning
- Cleaning abrasive too coarse

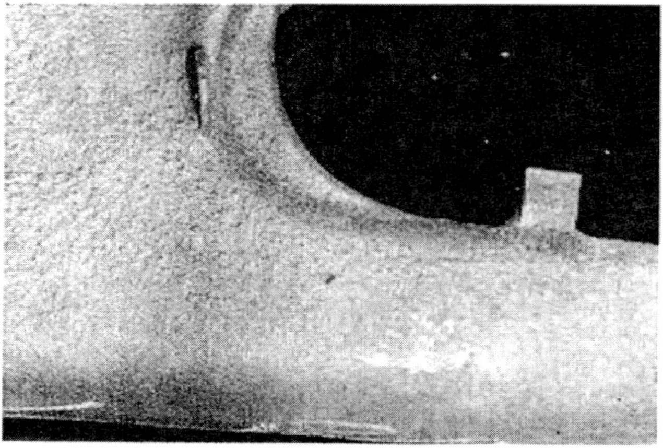

Fig. 20-5. Incorrect choice of molding sand caused excessive permeability in this casting.

Fig. 20-6. A sand content high in limestone created this rough effect (pocket) in the areas of maximum temperature.

CHAPTER 21

Runouts and Bleeders

Description

Runouts and bleeders are defects which result in an incomplete casting. Runouts occur during pouring and usually the mold cavity has not been completely filled. Bleeders occur after pouring has stopped and the mold cavity has been filled.

Fig. 21-1. A typical incomplete casting caused by runout.

Causes

1. Casting and Pattern Design

- None

2. Pattern Equipment

- A pattern mounted on undersized board or plate
- Worn bearing surfaces
- Dirty bearing surfaces on plate or flask
- Mismatched patterns
- Thin pattern plate

3. Flask Equipment and Handling Devices

- Flask size too small
- Warped or twisted flask
- Dirty flask joint
- Faulty pins and bushings
- Worn hinges and locks on snap flasks
- Warped and twisted bands
- In a cope and drag arrangement the pattern plate is mounted higher than the flask landing strip
- Weak, burned or uneven bottom boards
- Improperly fitting jackets
- Insufficient or improperly placed weights
- Bumping molds on a conveyor
- Insufficient drag depth

Fig. 21-2. A weak core permitted metal to run into this core cavity (it was impossible to remove this core section).

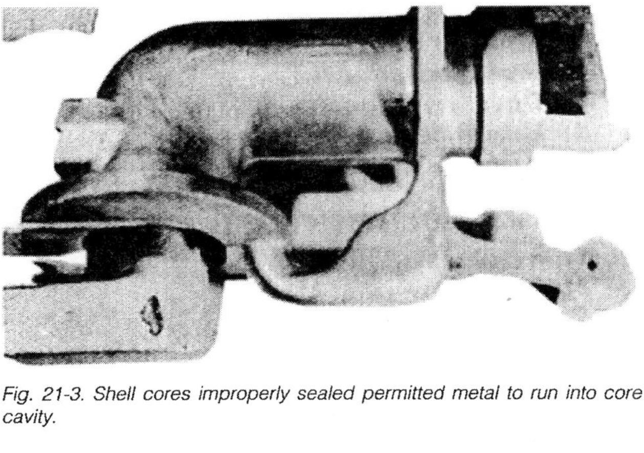

Fig. 21-3. Shell cores improperly sealed permitted metal to run into core cavity.

Fig. 21-4. A poorly bedded bottom board permitted this casting to bleed during pouring.

4. Gating and Risering

- Sprue, runner or riser located too close to the flask
- Patterns located too close to the flask
- Excess head pressure

5. Molding Sand

- Weak sand
- Low green deformation in sand
- Excessive expansion and contraction in shell molds
- Late gas explosion in the mold

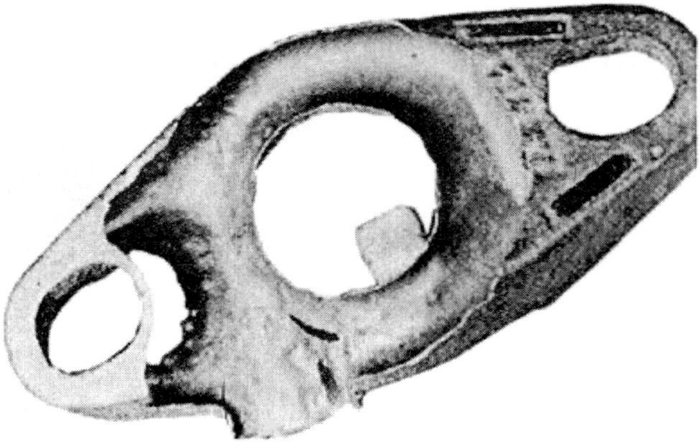

Fig. 21-5. This bleeding was caused by shifting of the jackets before solidification while the mold was on a moving conveyor. Note that the mold had been full.

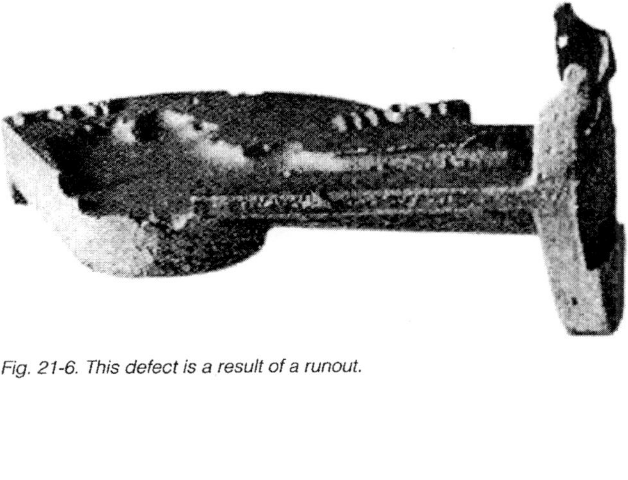

Fig. 21-6. This defect is a result of a runout.

6. Core Practice

- Coreprints too large
- Vents located too close to surface or not properly sealed
- Weak cores
- Shell cores with thin spots
- Shell cores not sealed at prints
- Thermal cracking of shell cores
- Excessive peel back (shell cores)
- Too rapid heating or cooling
- Resin strength too low for application

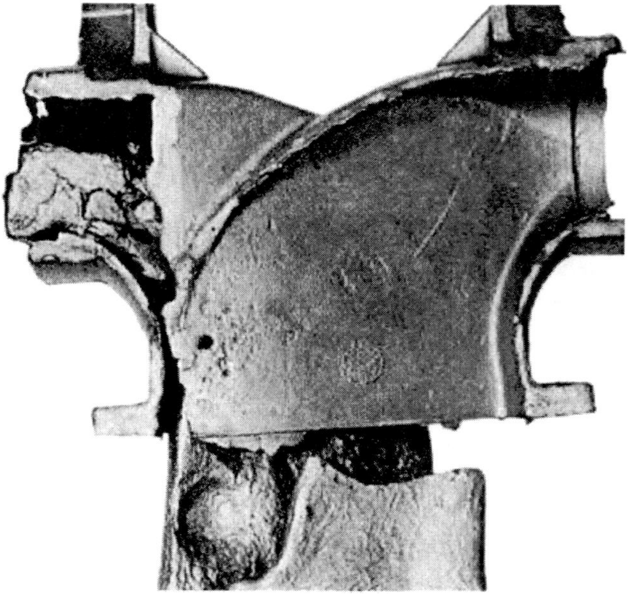

Fig. 21-7. This runout condition was caused by breakthrough in a thin area of the shell core.

7. Molding Practice

- Soft ramming
- Insufficient or improperly weighted molds
- Uneven clamping
- Improperly bedded bottom boards
- Non-uniform hardness or lumpy bedding sand
- Careless mold closure
- Venting located too close to the mold cavity or not sealed properly
- Lack of a proper cope seal in large molds
- Under or over baked dry sand molds
- Heating of skin dried molds too rapid
- Insufficient reinforcing or support of dry sand molds

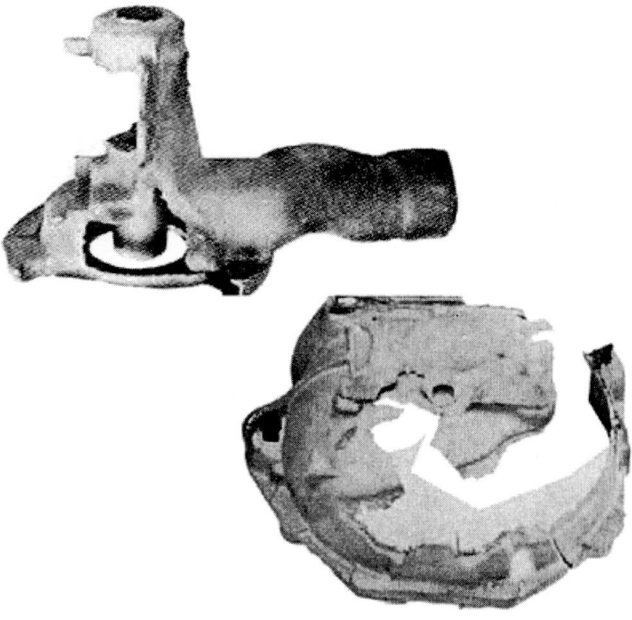

Fig. 21-8. These runouts emptied much of the mold cavity (above and below).

8. Metal Composition

- High fluidity
- Wide solidification range

9. Melting Practice

- None

10. Pouring Practice

- Failure to use jackets, weights or clamps
- Pouring too hard or high
- Pouring too hot
- Bumping molds during pouring
- Removing weights, clamps or jackets too soon

11. Miscellaneous

- Shakeout too hot
- Knocking off gates and risers too soon

CHAPTER 22

Scars, Seams and Plates

Description

A scar is a minor or shallow shiny mark on a casting surface where the casting does not conform to the pattern. A seam is seen as an indented line on the casting surface. A plate is a layer of metal partially separated from the main body of the casting section. All of these defects usually occur on flat surfaces.

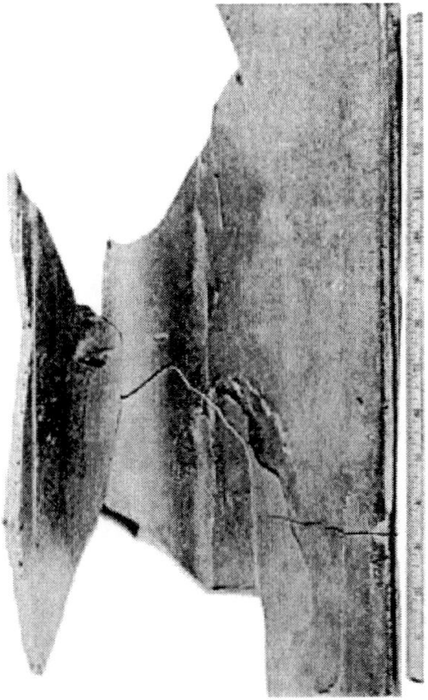

Fig. 22-1. Plate defect caused by wet sand and overly hard ramming. The molten metal forced out by gas pressure later flowed over previously solidified metal.

Causes

1. Casting and Pattern Design

- Abrupt change in casting section, resulting in interrupted metal flow
- Sharp corners resulting in interrupted metal flow

Fig. 22-2. This figure illustrates a typical plate defect.

Fig. 22-3. These plates were caused by a gas defect.

2. Pattern Equipment

- Worn patterns or coreboxes causing thin metal sections
- Incorrect pattern dimensions resulting in casting sections too thin for the fluidity of the metal
- Patterns were improperly reinforced causing excessively thin sections

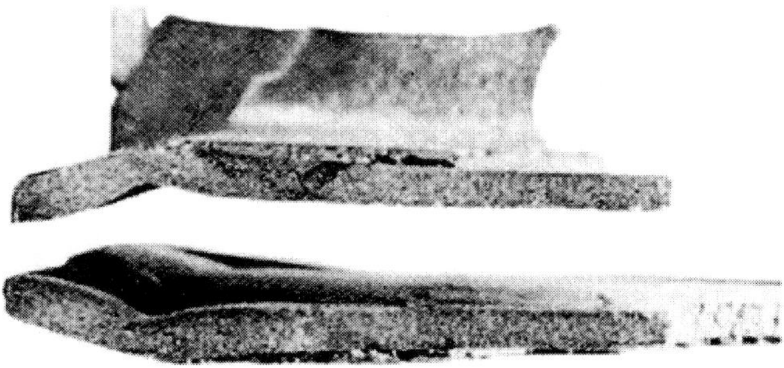

Fig. 22-4. A cross section of a plate defect caused by wet sand and hard ramming.

3. Flask Equipment and Handling Devices

- Cope and drag shift caused thin metal sections
- Pattern improperly mounted
- Lack of tilting arrangement for thin flat sections
- Flask insufficiently barred allowing cope to sag
- Weak bottom boards

4. Gating and Risering

- Gate design created turbulent flow
- Gate design created interrupted flow
- Gate design promoted slow pouring
- Low head pressure
- Widely separated or too few gates

5. Molding Sand

- Low permeability
- Excessive free water
- Gas forming material content too high
- Differences in strength or permeability of facing and backup sand
- High expanding sand
- Low hot deformation

6. Core Practice

- Low permeability from excessive fines and poor grain distribution
- Cores too high in gas forming material
- Insufficient or plugged vents
- Undercured cores
- Off size or distorted cores
- Rubbed core surface
- Broken or thin wash coating

7. Molding Practice

- Combination of hot and cold materials
- Non-uniform ramming
- Insufficient venting
- Low head pressure
- Thin metal section
- Excessive sticking and patching
- Turbulent metal flow
- Insufficient size or number of chaplets
- Gas pockets in thin sections

8. Metal Composition

- Low fluidity
- Too low carbon equivalent
- Excessive carbide stabilizers
- Excessive ladle additions

9. Melting Practice

- Cold melting
- Gassy metal
- Improper flux treatment
- Incorrect composition
- Excessive ladle additions

10. Pouring Practice

- Interrupted pouring
- Lack of synchronization in pouring multiple down sprues
- Pouring too high
- Reducing rate of pouring too soon
- Thin flat sections poured level instead of tilted
- Excessive mold weights
- Rough handling
- Inferior quality jackets
- Rough handling of molds

11. Miscellaneous

- Dirty, oxidized or damp chills, chaplets or nails
- Excessive coating on chills

Fig. 22-5. A gray iron cylinder liner with a seam defect caused by an interrupted pour is shown.

CHAPTER 23

Shifts

Description

A shift is a mismatch of the casting at the parting line. Shifts may occur in cores or core assemblies.

Fig. 23-1. A typical mold shift caused by a mismatched cope and drag.

Causes

1. Casting and Pattern Design

- Inadequate provision for coreprint or core support

2. Pattern Equipment

- Pattern not properly aligned in the cope and drag
- Misaligned locators on non-shift devices
- Loose or worn dowel pins in loose patterns
- Patterns not properly secured to plates
- Lack of locators
- Inadequate or improper coreprints

3. Flask Equipment and Handling Devices

- Worn, bent or missing pins and bushings
- Tight, worn, bent or broken bands and jackets
- Weak or warped flasks
- Faulty or worn machines
- Worn setting, assembling, rubbing or checking jigs
- Inadequate space between patterns or patterns and flask

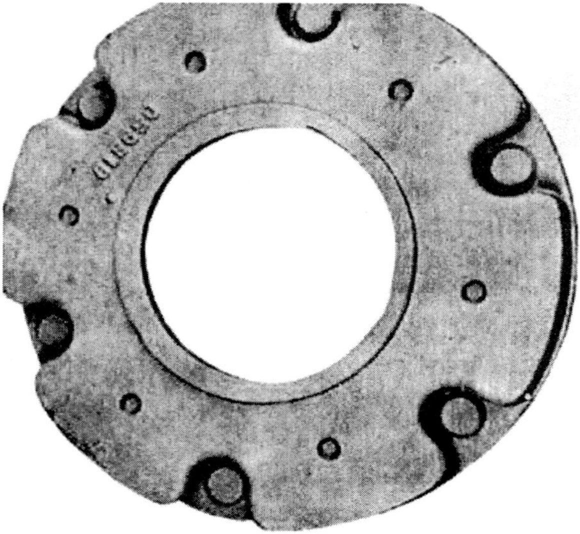

Fig. 23-2. Shift due to poor alignment of cope and drag patterns.

Fig. 23-3. Two castings showing the result of inadequate coreprints, producing a vertical shift (defective casting is located on left).

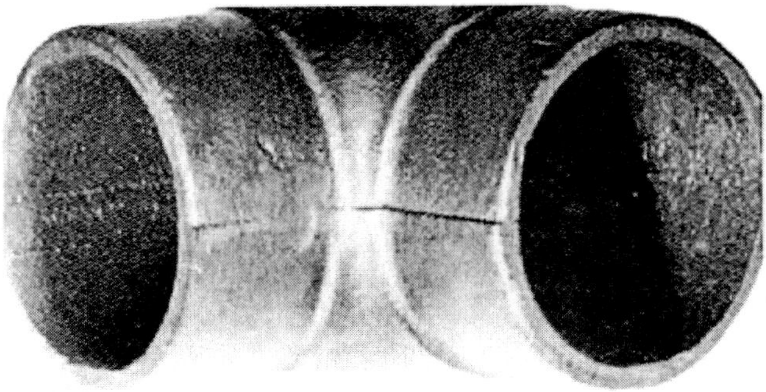

Fig. 23-4. Worn pins and bushings cause a mismatch in one or more directions.

Fig. 23-5. A defective casting as a result of incorrect core assembly is viewed on the left, the new casting after correction on the right.

4. Gating and Risering

- None

5. Molding Sand

- Weak sand

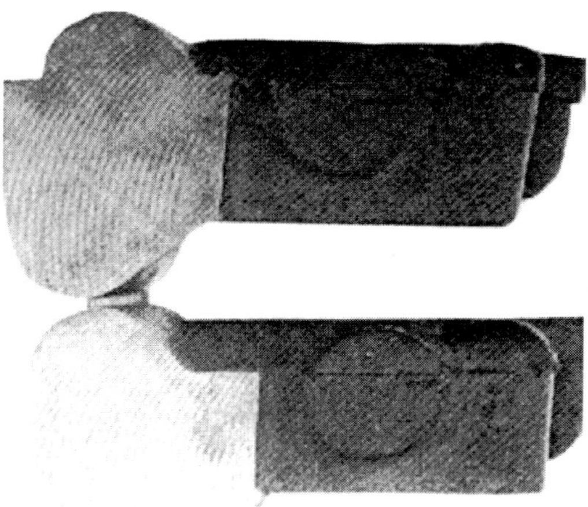

Fig. 23-6. Shift as a result of cope motion during transport of the mold to the pouring floor

6. Core Practice

- Improperly aligned cores in assembly
- Excessive core paste
- Core paste not dry
- Excessive rubbing of core
- Distorted core

7. Molding Practice

- Excessive rapping of loose patterns
- Use of dowel pin holes for rapping holes
- Improper placement or unsecured chaplets
- Incorrect chaplet: Too long, too short, not heavy enough to support load
- Core set wrong
- Reversing cope or drag
- Improper or careless clamping
- Improper or careless placement of mold weights
- Rough or improper handling

  Molder carrying mold against his body

  Bumping on a pallet line

  Bumping in automatic handling

  Rough roller conveyor

  Tilting or upending molds
- Vertical Molding

  Sand buildup on the rails may cause twisting of the mold or jumping, causing the mold faces to move opposite of one another.

  Improperly adjusted or aligned side clamps

  Improperly aligned stationary rail or moving rail sections

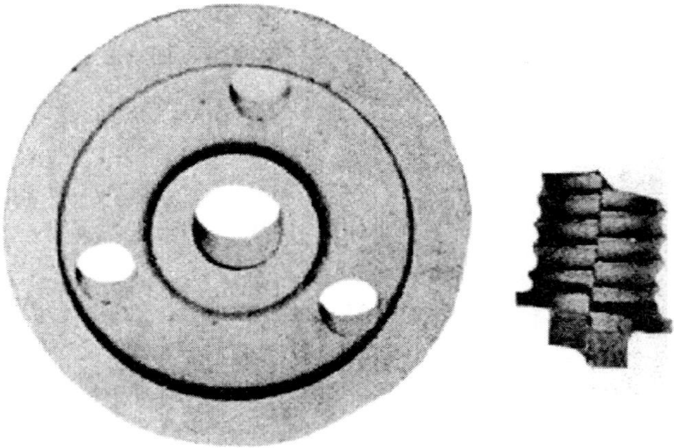

Fig. 23-7. Two examples of cope and drag shift due to improper handling of mold jackets.

8. Metal Composition

- None

9. Melting Practice

- None

10. Pouring Practice

- Improper handling of weight and jackets
- Bumping molds with pouring ladle

11. Miscellaneous

- Excessive vibration

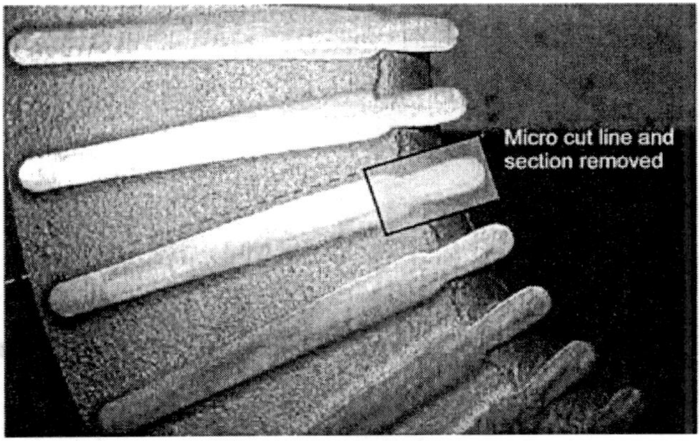

Fig. 23-8. Cut away view of a brake disc rotor showing the mismatch on the cooling fin resulting from misalignment between the stationary rails and the moving rails for a vertically parted mold.

Fig. 23-9. Metallographic sample from the area designated above showing the mismatch.

Fig. 23-10. Another example of a casting with mold shift is shown above.

CHAPTER 24

Shot Metal or Cold Shots

Description

Shot metal or cold shots are small globules of metal imbedded in, but not entirely fused with the casting. A cold shot appears as nearly round shot, while a cold shut is a form of a misrun and appears as a crack or seam.

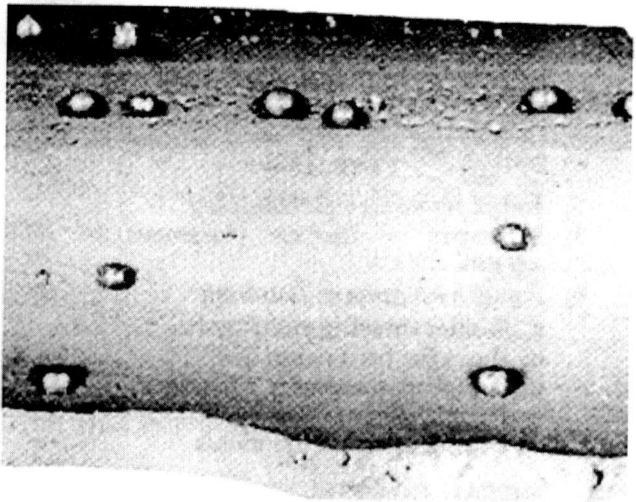

Fig. 24-1. Cold shot on the inside surface of a centrifugally-cast iron pipe. This particular defect was eliminated by ladle additions of silicon.

Causes

1. Casting and Pattern Design

- Improper design features:
  Sharp corners
  No fillets/small fillets
  Flow divisions, turns
- Dead cavities

2. Pattern Equipment

- Too much clearance on prints requiring excessive amounts of sealing compound
- Inadequate venting at the parting line
- Any gating design that results in turbulence

3. Flask Equipment and Handling Devices

- Bars and flask located too close to the mold cavity surface
- Sprues located too close to bars or flask walls

4. Gating and Risering

- Gating design that causes turbulence (converging metal flows) or interrupted flow
- Multiple gating that results in the majority of the metal entering into one or two ingates
- Pop or pencil gates that are too small.

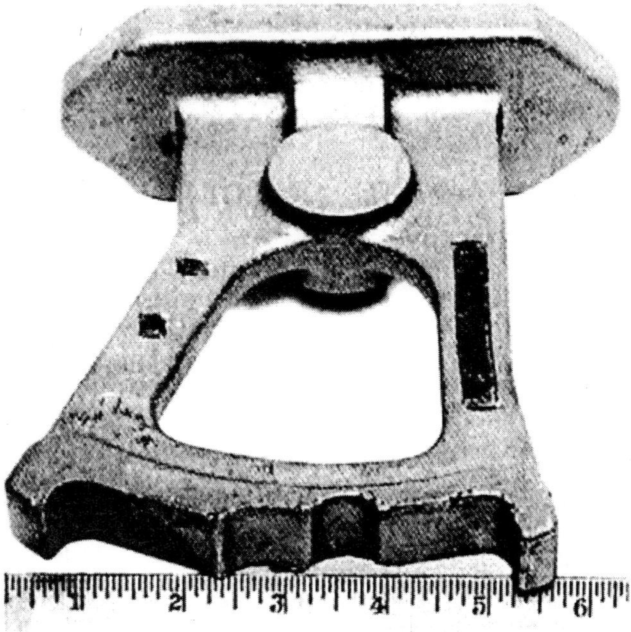

Fig. 24-2. Cold shot iron as a result of interrupted pouring.

5. Molding Sand

- High moisture or excessive gas that results in a severe blowing action causing the metal to spatter
- Foreign material (such as clay balls, etc.) that produces excessive gas in a localized area
- Improper mixing leads to an uneven distribution of water/additives, especially gaseous additives

6. Core Practice

- An undercured core or excessive gas from an undercured core, gaseous binders, additives, core wash, or core paste
- Exposed wires, arbors, and supports that are wet, dirty, oily, or rusty

7. Molding Practice

- Poorly dried molds
- Plugged vents
- Excessive sticking of mold

8. Metal Composition

- Certain metals that are more prone to rapid oxidation prevents the shot from redissolving (i.e., low carbon steel or aluminum)
- Off analysis metal that results in an excessive solidification rate
- Phosphorus levels too high (centrifugal casting)

9. Melting Practice

- Oxidized metal
- Violation of melting process (too low melt temperature, too short melt process)

10. Pouring Practice

- Careless skimming, spilling or splashing metal in a sprue or riser
- Interrupted pouring
- Cold metal
- Pouring too fast or too slow
- Damaged ladle or crucible tip
- Any foreign material which causes metal to splatter during pouring

11. Miscellaneous

- Condensation in the mold
- Long delay before pouring; Sand core or coating acquires moisture
- High ambient humidity
- Very low ambient temperature
- Cold metal (permanent mold, centrifugal casting)

CHAPTER 25

Shrinkage Cavities and Depressions

Description

A shrinkage cavity is a jagged hole or spongy area lined with fern-like structure called dendrites. A shrinkage depression on the surface of the casting is a recession from the true plane of the mold surface. There is apt to be confusion determining between a shrink and a blow. If a doubt exists, refer to the chapter on *Gas Defects*. Many of the cures for both are the same, but in some cases the cure is just the opposite. Shrink characteristically appears in heavier sections, at abrupt changes of section thickness or at hot spots. These same locations are also prone to producing gas defects and hot tears.

Causes

1. Casting and Pattern Design

- Abrupt change in section size
- Isolated heavy sections which cannot be fed
- Inadequate fillets/too small fillets
- Insufficient area for feeding or chilling isolated sections

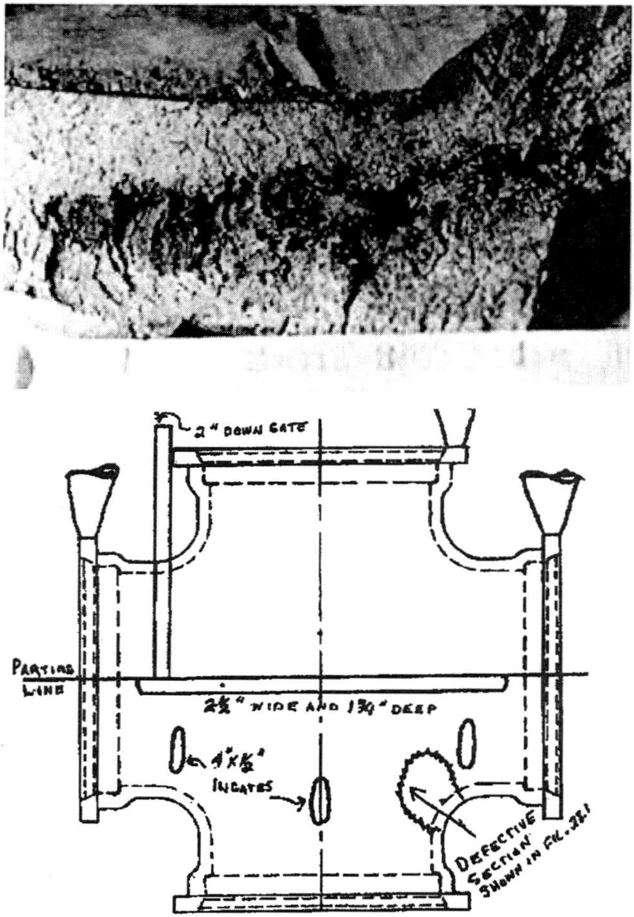

Fig. 25-1. A typical shrinkage defect in a thick section that was inadequately fed (casting design shown in schematic).

2. Pattern Equipment

- Worn patterns and/or coreboxes
- Incorrect pattern dimensions resulting in thin or heavy sections
- Poor pattern construction, improperly reinforced for high-pressure molding

3. Flask Equipment and Handling Devices

- Insufficient cope height
- Flask too small
- Improperly fitted jackets
- Too many patterns in flask

4. Gating and Risering

- Gates and risers not promoting progressive solidification
- Insufficient number and/or improperly attached gates and risers
- Risers too small
- Improper ratio of riser neck to casting size
- Large ingates
- Riser improperly located
- Riser contact area is an unacceptable shape or too small
- Riser improper shape
- Wrong metal feeding to riser
- Poorly insulated riser cavity

5. Molding Sand

- Low green compressive strength
- High green deformation
- High hot deformation
- Wet sand

6. Core Practice

- Core movement-oversize prints
- Insufficient chaplets
- Weak cores
- Soft or insufficient prints
- Lack of fillets

7. Molding Practice

- Any molding practice that allows mold wall movement under pressure-refer to molding sand
- Omission of exothermic or insulating materials when required
- Lack of chills or incorrect chill location
- Soft ramming

Fig. 25-2. Insufficient cope height prevented adequate riser height to feed sections involved. Sectioning through both the risers and the casting demonstrates the need for additional feed metal height.

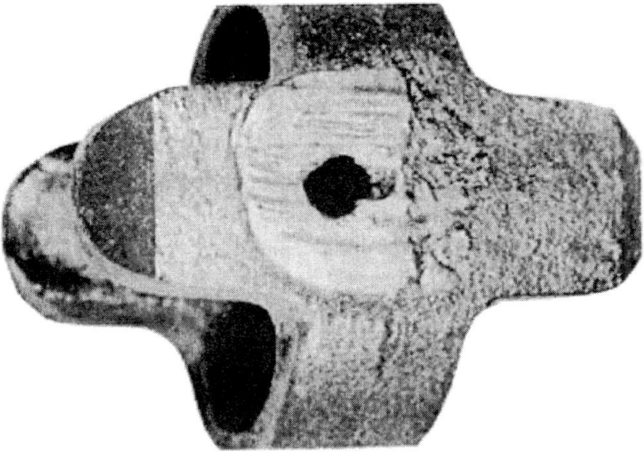

Fig. 25-3. Feeding four of these castings from one small riser was not adequate enough to prevent shrink in the riser neck.

Fig. 25-4. Gate area was equal to casting section creating a shrink in the gate.

8. Metal Composition

- Chemical composition unsuited for section involved
- Low carbon content
- Ductile iron

  Too high/too low carbon content

  Too high/too low carbon equivalent

  High silicon

  High nodule counts

  Increase in thermal analysis parameters TAL (Thermal Arrest Liquidus), TER (Thermal Eutectic Recalescence) and TEU (Thermal Eutectic Undercooling)

  Increased magnesium content

Fig. 25-5. A soft ram mold allowed mold-wall movement thus creating a feeding problem in the heavier section of the casting.

9. Melting Practice

- Careless charging or weighing
- Incorrect composition
- High temperature metal
- Increased oxidation

10. Pouring Practice

- Pouring too hot and/or too cold
- Failure to top off risers with hot metal

Fig. 25-6. The surface deflections (arrows) are probably due to inadequate gating/feeding or risering of the casting.

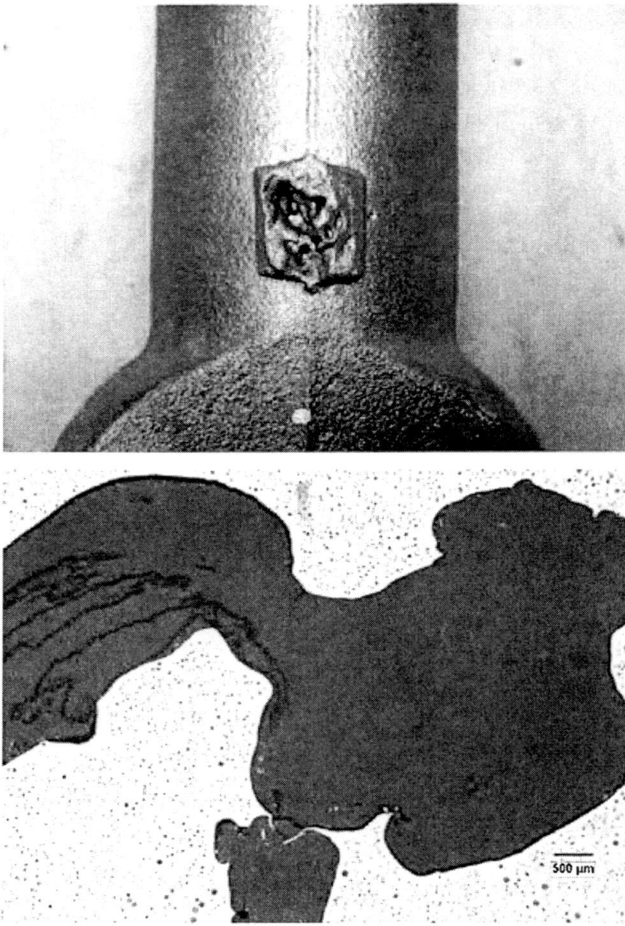

Fig. 25-7. A shrinkage cavity located in the gate area and the corresponding cross section of the cavity located on right.

Fig. 25-8. This shrinkage is located at the thermal center of the casting.

11. Miscellaneous

- Cracked molds
- Run outs
- Run in (shell cores)
- Improperly clamped molds

CHAPTER 26

Stickers

Description

A sticker is excess metal on the surface of the casting caused by a portion of the mold face remaining on the pattern.

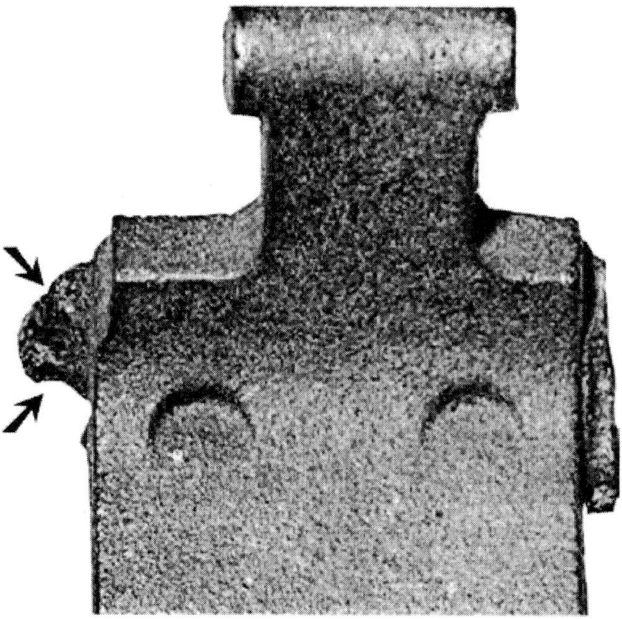

Fig. 26-1. A typical lump on a casting surface caused by sand adhering to the pattern during molding is seen protruding on the left side of the casting.

Causes

1. Casting and Pattern Design

- Insufficient draft

2. Pattern Equipment

- Poorly polished or improperly coated
- Worn or rough
- Fillets and depressions filled with wax
- Insufficient draft or back draft
- A pattern undercut or loose at parting line
- Failure to vent critical pockets

3. Flask Equipment and Handling Devices

- A flask lacking rigidity
- Insufficient flask height
- Too many patterns for the flask size
- A worn or misaligned pattern draw device
- Over jolting, out-of-level or worn molding machine
- Excessive or insufficient vibration
- An incorrectly adjusted automatic parting spray

4. Gating and Risering

- Improperly placed gates, runners and risers causing pockets

Fig. 26-2. Excessive moisture used in an attempt to compensate for hot sand caused general sticking of sand grains to the pattern surface.

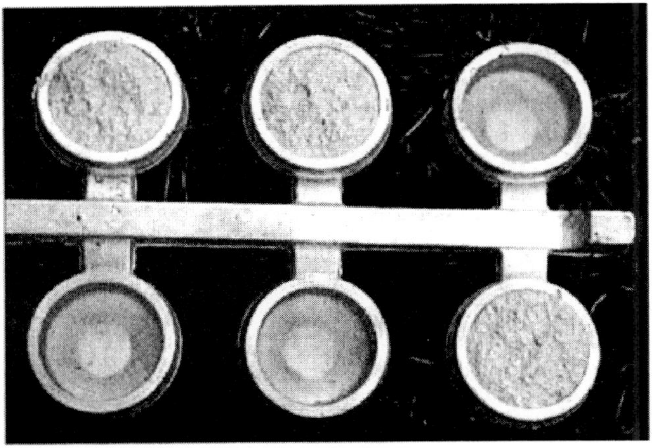

Fig. 26-3. Excess metal on the surface of these castings was caused by a portion of the mold face sticking to the pattern.

5. Molding Sand

- Hot sand
- Inadequate sand properties
  - Low green tensile strength
  - Low green shear strength
  - Low permeability
  - Poorly mixed sand
  - Wet sand

6. Core Practice

- An oversized ramup core
- Rough, dirty, or lack of draft in a core box
- Reaction between the binder and parting
- Excessive moisture in sand

7. Molding Practice

- Rough handling of the mold
- Excessive liquid parting
- Excessive or insufficient vibrating or rapping of the pattern
- Too great of a temperature difference between the pattern and the sand
- Leaving the pattern in the mold too long
- Ramming directly on the pattern with a slinger
- Insufficient mold reinforcement
- Gaggers, nails, soldiers, bars located too close to the mold cavity
- Careless or improper drawing of the pattern

8. Metal Composition

- None

9. Melting Practice

- None

10. Pouring Practice

- None

11. Miscellaneous

- Foreign material

CHAPTER 27

Swells, Strains & Sags

Description

A swell is an enlarged metal section related to mold wall movement and the displacement of sand. A strain is a swell with a fin that splits the mold open. A sag is an increase or decrease in a metal section due to sagging of the cope or core. A core fin is a depression in the casting caused by a fin on the core which was not removed before the core was set.

Fig. 27-1. This typical swell relates to mold-wall movement and the displacement of sand.

CAUSES

1. Casting and Pattern Design

- A design requiring long or thin cores that prevents the use of a support
- Any design which contributes to soft ramming
- Large flat surfaces
- Large castings which require high metal pressure

2. Pattern Equipment

- Inaccurate parting lines
- Inaccurate coreprints
- Inaccurate dimensions (kiss-throughs)
- Warped coreboxes
- Worn pattern equipment
- Flask landings occur below the parting line.
- Warped, strained, or off-dimension matchplates
- An inadequately reinforced pattern
- Excessive number of multiple patterns
- Inadequate relief vents

Fig. 27-2. A strain is an enlarged metal section related to mold-wall movement (swell) with fins.

3. Flask Equipment and Handling Devices

- Warped, burned, broken, or weak bottom boards
- A carved or worn strike-off
- Weights that are too light for the application
- Worn or warped flasks
- Poorly fitting jackets
- Dirty jackets

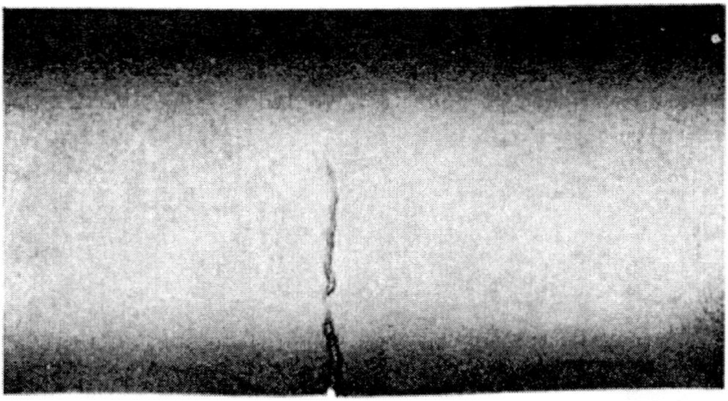

Fig. 27-3. A sag in the core created a thin section with a corresponding thick section.

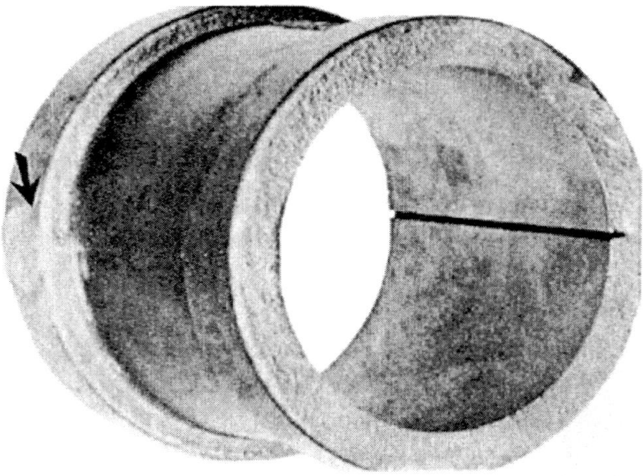

Fig. 27-4. This defect resulted from a fin that should have been removed from the core prior to setting.

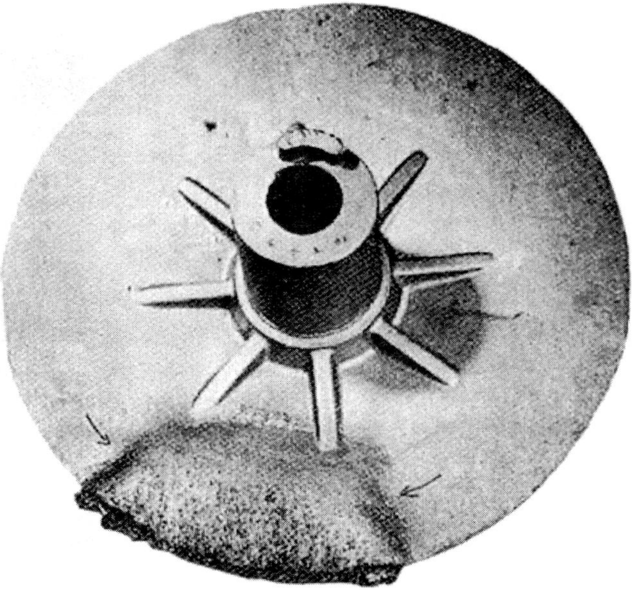

Fig. 27-5. Insufficient clearance between bars caused the molder to soft ram this section.

Fig. 27-6. A weak core distorted during pouring causing this sag.

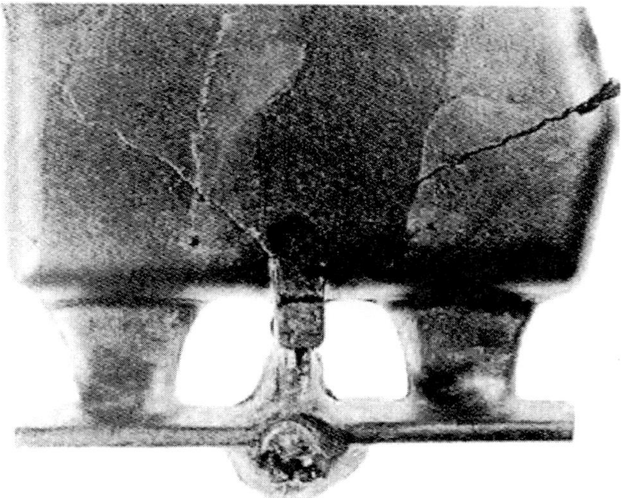

Fig. 27-7. A poorly cleaned bottom board strained the mold and caused this distortion.

4. Gating and Risering

- Lack of pop-offs or relief vents
- Inadequate bearing surface

5. Molding Sand

- Weak sand
- Any condition that permits excessive mold wall movement
- Mold develops a crack due to the expansion of sand

6. Core Practice

- Core too small for print
- Insufficiently reinforced cores
- Setting the wrong core
- Weak cores (improperly blown, rammed, or wrong mix)
- Cracked, broken or warped cores (twisted drier)
- Excessive hot deformation
- A thin section on shell core
- Core develops crack due to sand expansion

Fig. 27-8. A soft ram on the mold was aggravated by a hard pour, causing this swell at the parting line.

7. Molding Practice

- A soft ram, not tucked or properly peened
- Excessive shaving of the joint
- Poorly reinforced molds
- Uneven clamping of molds
- Ramoff or ramaway
- Uneven bedding of the bottom board

8. Metal Composition

- The high density metals have a greater tendency to produce mold-wall movement
- Pouring temperature
- Expansion and contraction (ductile iron)

9. Melting Practice

- Inoculation of graphitic metals increases mold-wall movement

10. Pouring Practice

- Ladle too high

11. Miscellaneous

- Dirty pallets
- Excessive spinning speeds in sand molds

CHAPTER 28

Veins and Fins

Description

Veining appears on the casting as an irregular fin (or fins) of metal protruding from the surface of the casting. Two conditions must exist for this type of defect to occur. The core or mold surface develops a crack either from expansion-contraction forces or mechanical handling and the metal must be fluid enough to enter the crack.

NOTE: (T) Signifies thermal cracking
(M) Signifies mechanical cracking

Fig. 28-1. This casting exhibits severe veining. This defect is also referred to as finning.

CAUSES

1. Casting and Pattern Design

- Sharp corners that develop hot spots (T)
- Thin cores surrounded by heavy metal sections (T)
- Massive sections, refer to *Chapter 27-Swells, Strains, Sags and Core Fins* (M)
- Pin or thin cores placed horizontally instead of vertically when possible (T)

2. Pattern Equipment

- Sharp corners that develop hot spots (T)
- Thin cores surrounded by heavy metal sections (T-M)
- Uneven metal sections causing uneven heat distribution in shell or hot box equipment (T)
- Inadequate support from core dryers causing cracked cores (M)
- Warped core dryers (M)
- Dryers not properly relieved (M)

3. Flask Equipment and Handling Devices

- Rough mechanical handling of uncured cores (M)
- Misaligned shell or hot box equipment cracking cores (M)
- Flasks not rigid enough for molding process involved (M)
- Loose bars
- Rough conveying of cores or molds (M)

4. Gating and Risering

- Gate causing metal to impinge on core causing hot spot in critical area (T)
- Riser too close to core print (hot spot) (T)
- Insufficient number of gates causing local hot spots (T)

Fig. 28-2. This test bar exhibiting severe veining was poured after soft ramming.

5. Molding Sand

- Excessive organic material in sand (seacoal, etc.) (T)
- Excess moisture in sand for skin or oven dried molds (T)
- Improper mulling (T-M)

6. Core Practice

- Weak cores (green or cured) (M)
- Cores too dense (T)
- Excess binders (T)
- Too rapid heating and cooling in core oven (T)
- Improperly reinforced cores-rods, wires, or bars (M)
- Inadequate dryer support-warped dryers (M)
- Baking temperature too high (T)
- Over-cured cores (T)
- Brittle cores (T-M)
- Improper portion of hexamethylene tetramine in phenol-formaldehyde shell cores for humidity conditions prevailing (T)
- Cores stored under conditions of high humidity (T)
- Poor sand mixing (T-M)
- Inadequate hot plasticity of cores (T)
- Insufficient iron oxide content (T-M)

7. Molding Practice

- Excessive sticking of mold surface (T)
- Inadequately reinforced or supported molds (M)
- Non-uniform torching of skin dried molds (T)
- Uneven clamping of molds (M)
- Excessive mold wash (T)
- See "*Section 3, Flask Equipment and Handling Devices*" for additional factors which control the molding aspect of these defects.

8. Metal Composition

- High carbon-silicon-phosphorus content in gray irons (T-M)

9. Melting Practice

- None

10. Pouring Practice

- Pouring too hot for metal composition involved (T)
- Pouring too hard causing metal to forcibly impinge against a core (ladle too high) (T)

11. Miscellaneous

- In centrifugal casting, any condition that causes a lack of adherence of the sand to metal mold

Fig. 28-3. This gray iron casting exhibits severe veining due to a high percentage of fines in the molding sand mixture.

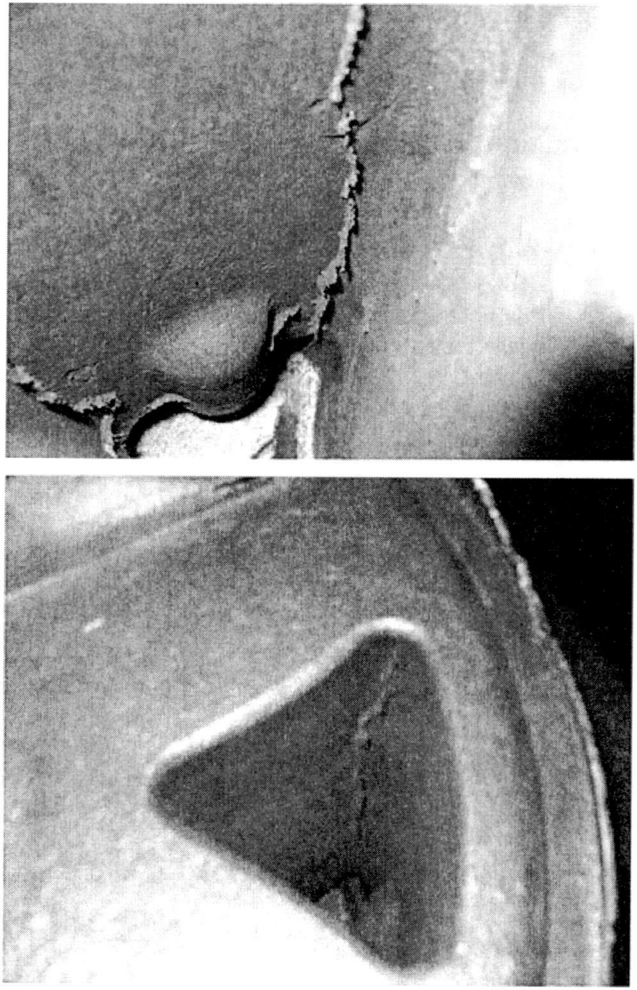

Fig. 28-4. Two different views of a casting with veining defects are shown.

CHAPTER 29

Warped Castings

Description

Warpage is a deformation that occurs during or after solidification, mostly in gray or white cast iron (non-ductile). The problems are the same as those described in *Chapter 11 Hot Tears*, namely differential stresses set up during cooling or during the processing of the castings. The difference between the defects (aside from appearance) is either in the degree of stress or difference in the tearing tendency of various metals. For example, a stress situation that would hot tear a steel or malleable casting might only warp a gray or ductile iron casting.

CAUSES

1. Casting and Pattern Design

- Large flat uninterrupted sections
- Lack of ribs

2. Pattern Equipment

- Large flat sections can be broken up with wavy or roughened sections provided this does not interfere with the application (including appearance) of the casting.
- Pattern not adapted to compensate for warping
- Incorrect number of flask tie bars

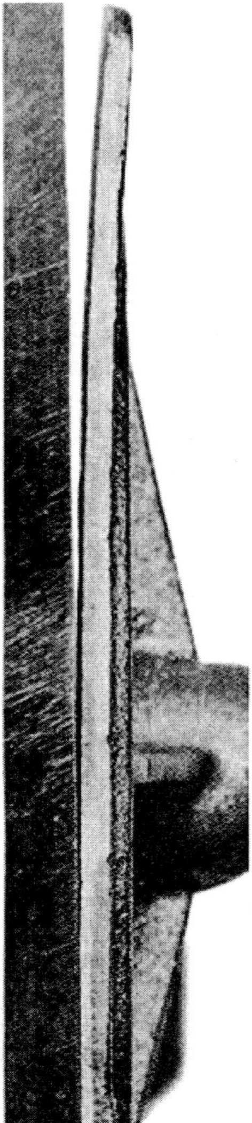

Fig. 29-1. An incorrect design increased the tendency of this casting to warp.

Fig. 29-2. The casting on the left shows a typical warp when compared to the normal malleable casting on the right.

3. Flask Equipment and Handling Devices

- Flask tie bars located too close to pattern
- Flask tie bars located too close to risers and downsprues
- Weak bottom boards
- Flask too small
- Weak or worn flasks
- Weak or deteriorated squeeze board
- Inappropriately adjusted squeeze head, including misalignment

4. Gating and Risering

- Gating system which prevents the desired contraction or casting shrinkage upon cooling
- Gate area too small
- Improper location of ingates

5. Molding Sand

- Green strength too low
- Flowability or moldability too low
- Hot strength too low or too high
- Poor compactability resulting in low and non-uniform mold hardness

6. Core Practice

- Poor collapsibility of cores
- Hot strength too high
- Over-reinforced core (rods and arbors)
- Rods located too close to casting surface

7. Molding Practice

- Improper relationship of mold hardness between the cope and drag. Soft cope allowing upward movement of the contracting metal or vice versa

  Warpage can be alleviated or reduced by ramming or compacting the area of the mold harder in the direction of the warpage. After pouring, the mold can also be rammed by hand or pneumatically.

8. Metal Composition

- High combined carbon
- Excessive use of carbide stabilizers
- Chills etc., which promote the stabilization of combined carbon.

9. Melting Practice

- Careless charging produces a wide range of iron chemistries
- Metal too cold for proper dissolution of additions or trim alloy
- Any metal practice that stabilizes carbides or changes metal composition

10. Pouring Practice

- Pouring too hot or cold for the casting design

11. Miscellaneous

- Shakeout too hot
- Castings not properly supported during heat treatment
- Peening action during cleaning, especially on large flat area of the casting
- Excessive burn-in requires longer cleaning cycles
- Machining may create a partial stress relief
- Large castings not properly supported during shipment

CHAPTER 30

Widmanstattën Graphite

Description

Widmanstattën graphite occurs in the microstructure when contamination from heavy metal occurs. The principal element that causes this deleterious structure is lead. The presence of this graphite structure can decrease the mechanical properties by as much as 50%–70% in either gray iron or ductile iron. Widmanstattën means precipitation on crystallographic planes. See the three figures below for examples of graphite precipitation after solidification at varying magnifications.

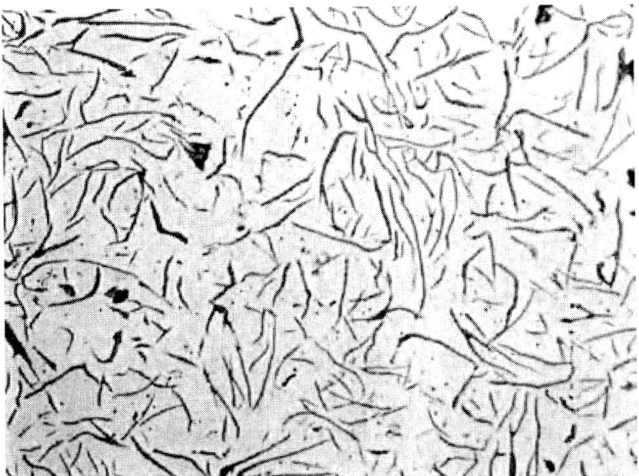

Fig. 30-1. Graphite precipitation after solidification is shown at 100X magnification.

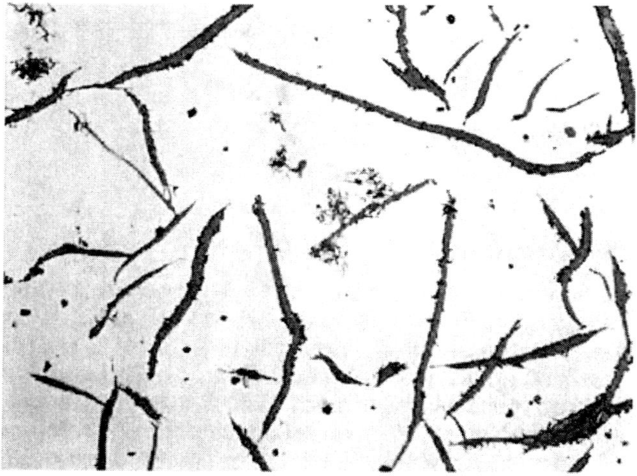

Fig. 30-2. Graphite precipitation after solidification is shown at 400X magnification. (Same casting as in Fig. 30-1.)

Fig. 30-3. Graphite precipitation after solidification is shown at 1000X magnification. (Same casting as in Fig. 30-1.)

Causes

1. Casting and Pattern Design

- None

2. Pattern Equipment

- None

3. Flask Equipment and Handling Devices

- None

4. Gating and Risering

- None

5. Molding Sand

- None

6. Core Practice

- None

7. Molding Practice

- None

8. Metal Composition

- Lead content in excess of 5ppm (0.0005%).

9. Melting Practice

- Contamination originated from scrap supplied with leaded steel
- Contamination from an unknown source with solder
- Contamination from lead originally used as a weight or ballast that was undetected in scrapped components

10. Pouring Practice

- None

11. Miscellaneous

- None

APPENDIX

International Defect Codes/ Corresponding Chapter

International Defect Code	Defect Name	CDHB Chapter
A112-GI	Veining or Finning	28
A114-GI	Fillet Scab	6
A115-GI	Fillet Vein	28
A121-GI	Cope Raised, Raised Mold	21
A122-GI	Sag or Strain	29
A211-GI	Swells	27
A212-GI	Erosion, Cut or Wash	3
A213-GI	Crush	2
A221-GI	Sticker	26
A224-GI	Mold Drop	5
A225-GI	Corner Scab	6
A226-GI	Broken or Crushed Core	2
B111-GI	Blowholes, Pinholes	8
B112-GI	Blowholes next to Inserts/Chills	8
B113-GI	Slag Blowholes	4, 8
B122-GI	Draws	8
B123-GI	Surface Pinholes	8
B124-GI	Dispersed Shrinkage	25
B211-GI	Open or External Shrinkage	25
B212-GI	Corner or Fillet Shrinkage	25
B213-GI	Core Shrink	25
B221-GI	Internal Shrinkage	25
C111-GI	Breakage	1
C121-GI	Hot Cracking	11
C221-GI	Hot Tearing	11
C311-GI	Cold Shut or Cold Lap	15
C331-GI	Cold Chill or Insert	10
D111-GI	Surface Folds	15
D112-GI	Cope Defect, Elephant Skin	7
D114-GI	Flow Marks	15
D121-GI	Rough Casting Surface	20
D131-GI	Buckle	7

D132-GI	Rat Tail	7
D133-GI	Flow Marks, Crow's Feet	20
D134-GI	Orange Peel, Alligator Skin	20
D141-GI	Sink Marks, Draw, Suck In	25
D142-GI	Slag Inclusions	4
D211-GI	Push Up, Clamp Off	2
D221-GI	Burn On	14
D222-GI	Burn In	14
D223-GI	Metal Penetration	14
D224-GI	Coating Failure	6
D231-GI	Scab, Expansion Scab	6
D232-GI	Cope Erosion Scab	6
E111-GI	Misrun	15
E122-GI	Poured Short	18
E123-GI	Runout	21
E221-GI	Broken Casting	1
F221-GI	Shift	23
F222-GI	Shifted Core	23
F223-GI	Ram Off, Ram Away	19
F233-GI	Casting Distortion	27
F234-GI	Warped Casting	29
G111-GI	Metallic Inclusions	4
G112-GI	Cold Shot	24
G121-GI	Slag Inclusions, Dross, Ceroxide	4
G131-GI	Sand Inclusions	4
G143-GI	Lustrous Carbon Films	15
G144-GI	Hard Spots	10
G211-GI	Chill	10
G221-GI	Type C Graphite	8, 17
G222-GI	Unacceptable Skin	20
G223-GI	Localized Hard Spots/Chilled Areas	10
G262-GI	Kish Graphite Inclusions	9
F	Off Dimension	16
B	Gas Voids with MnS	8
B	Gas Voids Lined with Graphite	8
B	Gas Voids Lined with Oxide	8
G	Micro-voids Lined with Steadite	8, 25
G	Tellurium Wash Migration	4
G	Carbide/Inverse Chill	12
G	Carbide-Intercellular Primary	10
G	Carbide Intercellular alloy	10
G	Widmanstattën Graphite	30